ABSTRACT

The forests of northern Wisconsin, a defining feature of the region's landscape, are expected to undergo numerous changes in response to the changing climate. This document provides a collection of resources designed to help forest managers incorporate climate change considerations into management and devise adaptation tactics. It was developed in northern Wisconsin as part of the Northwoods Climate Change Response Framework project and contains information from assessments, partnership efforts, workshops, and collaborative work between scientists and managers. The four interrelated chapters include: (1) a description of the overarching Climate Change Response Framework, a landscape-scale conservation approach also being expanded to other landscapes; (2) a "menu" of adaptation strategies and approaches that are directly relevant to forests in northern Wisconsin; (3) a workbook process to help incorporate climate change considerations into forest management planning and to assist land managers in developing ground-level climate adaptation tactics for forest ecosystems; and (4) two illustrations that provide examples of how these resources can be used in real-world situations. The ideas, tools, and resources presented in the different chapters are intended to inform and support the existing decisionmaking processes of multiple organizations with diverse management goals.

Cover Photo

A forest containing red pine and red oak trees. Photo by Maria K. Janowiak, U.S. Forest Service and Northern Institute of Applied Climate Science.

Manuscript received for publication September 2011

Published by:

USDA FOREST SERVICE
11 CAMPUS BLVD., SUITE 200
NEWTOWN SQUARE, PA 19073-3294

May 2012

For additional copies:

USDA Forest Service
Publications Distribution
359 Main Road
Delaware, OH 43015-8640
Fax: 740-368-0152

Visit our homepage at: **http://www.nrs.fs.fed.us/**

Forest Adaptation Resources: Climate Change Tools and Approaches for Land Managers

Edited by

Chris Swanston and

Maria Janowiak

One of many recreational lakes in northern Wisconsin.

Page intentionally left blank

AUTHORS

Chris Swanston is a research ecologist with the U.S. Forest Service, Northern Research Station and director of the Northern Institute of Applied Climate Science, 410 MacInnes Drive, Houghton, MI 49931, cswanston@fs.fed.us

Maria Janowiak is a climate change adaptation and carbon management scientist with the Northern Institute of Applied Climate Science, U.S. Forest Service, 410 MacInnes Drive, Houghton, MI 49931, mjanowiak02@fs.fed.us

Patricia Butler is a climate change outreach specialist with the Northern Institute of Applied Climate Science, Michigan Technological University School of Forest Resources and Environmental Science, 1400 Townsend Drive, Houghton, MI 49931, prbutler@mtu.edu

Linda Parker is a forest ecologist with the U.S. Forest Service, Chequamegon-Nicolet National Forest, 1170 4th Avenue So., Park Falls, WI 54552, lrparker@fs.fed.us

Matt St. Pierre is a biologist with the U.S. Forest Service, Chequamegon-Nicolet National Forest, 500 Hanson Lake Road, Rhinelander, WI 54501, mstpierre@fs.fed.us

Leslie Brandt is a climate change specialist with the Northern Institute of Applied Climate Science, U.S. Forest Service, 1992 Folwell Avenue, St. Paul, MN 55108, lbrandt@fs.fed.us

PREFACE

Forest Adaptation Resources: Climate Change Tools and Approaches for Land Managers is intended to provide perspective, information, and tools to land managers considering how to adapt forest ecosystems in northern Wisconsin to a changing climate. It describes a framework for responding to climate change that creates and gathers scientific information, establishes cross-boundary partnerships between ownerships and organizations, fosters close collaboration between scientists and land managers, builds useful tools that support diverse management goals, and finally seeks to deliver the fruits of these efforts in a timely and useful manner. It provides a wide-ranging "menu" of adaptation strategies and approaches relevant to northern Wisconsin and a "workbook" process to help land managers consider ecosystem vulnerabilities, select adaptation approaches that meet their needs, and devise tactics for implementing them. With the understanding that many of the resources in this document may initially seem abstract, this document also describes how the resources can be used in northern Wisconsin and provides examples developed in partnership with the Chequamegon-Nicolet National Forest. Although the focus is northern Wisconsin, we believe that the resources in this document are readily applicable to forest ecosystems throughout the Great Lakes, and broadly applicable to any forested landscape.

We seek to provide credible, relevant information through the *Ecosystem Vulnerability Assessment and Synthesis: A Report from the Climate Change Response Framework Project in Northern Wisconsin* (Swanston et al. 2011) and the Adaptation Strategies and Approaches (Chapter 2), and a process to help apply this information through the workbook (Chapter 3). Each step of the workbook process is driven by the user's goals, objectives, needs, and choices. This document does *not* make recommendations and is not intended to supersede existing decisionmaking processes. Our goal is to complement science-based management decisions made by multiple organizations, each with its own diverse goals, so that forest ecosystems across the landscape can better adapt to a changing climate.

Finally, this document is part of the Climate Change Response Framework Project, originally commissioned by the U.S. Forest Service Northern Research Station and Eastern Region. The Chequamegon-Nicolet National Forest, designated by the Eastern Region as a "Climate Change Model Forest for Landscape Management," has served as our test bed and major partner in the effort. The U.S. Forest Service Northeastern Area has been critical in reaching across institutional and ownership boundaries to engage the perspectives and expertise of a broad array of partners. The Wisconsin Department of Natural Resources and the Wisconsin Initiative on Climate Change Impacts have been integral to the entire Climate Change Response Framework. We extend our sincere appreciation to these organizations and the dozens of individuals who have contributed to this project.

Chris Swanston and Maria Janowiak, editors

Page intentionally left blank

CONTENTS

Page intentionally left blank

INTRODUCTION

The forests of northern Wisconsin are a defining feature of the region's landscape (Fig. 1), and these ecosystems, along with others, are expected to undergo many changes as a result of a changing climate (Swanston et al. 2011, Wisconsin Initiative on Climate Change Impacts [WICCI] 2011b). Wisconsin is already experiencing the effects of climate change, and many changes are projected over the next century (Kling et al. 2003; Kucharik et al. 2010; WICCI 2011a,b). Climate change is expected to have widespread effects on forest ecosystems in northern Wisconsin. Many of the important factors that influence forest composition and distribution are expected to change, including seasonal temperatures, the timing and type of precipitation, soil moisture patterns, the severity and frequency of natural disturbances, and the abundance of pests and diseases (Dale et al. 2001, Dukes et al. 2009, WICCI 2011b). Ecosystems are expected to respond to these changes in a variety of ways, often in reaction to increased stress (Swanston et al. 2011). While new research will continue to provide more detailed information about specific impacts, enough information is currently available to assess the vulnerabilities of regional forest ecosystems (Box 1; WICCI 2011b).

Figure 1.—Forests are an integral feature of the northern Wisconsin landscape.

Photo by Maria K. Janowiak, U.S. Forest Service and Northern Institute of Applied Climate Science

Box 1: Climate Change and Northern Wisconsin's Forests

Climate change will have many effects on the forest ecosystems of northern Wisconsin. An *Ecosystem Vulnerability Assessment and Synthesis* (Swanston et al. 2011) was developed to compile information on climate change effects and inform land managers in northern Wisconsin about the ecosystem components that are most vulnerable to climate change under a variety of future climate scenarios.

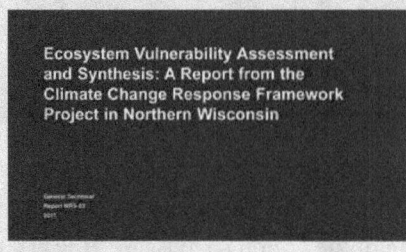

Ecosystem Vulnerability Assessment and Synthesis: A Report from the Climate Change Response Framework Project in Northern Wisconsin

The statements below are excerpts from the assessment and provide an overview of the major implications of climate change for these systems. Each excerpted statement also includes an assessment of the likelihood of occurrence using specific language established by the Intergovernmental Panel on Climate Change (IPCC) (2005). For more information, and to read each statement in its entirety, please consult the *Ecosystem Vulnerability Assessment and Synthesis: A Report from the Climate Change Response Framework Project in Northern Wisconsin* by Swanston et al. (2011).

Shifting Stressors

Climate change may relieve some stressors, while exacerbating others.

- **Temperatures will increase (virtually certain).** Annual increases in temperature represent the broadest possible stressor, strongly influencing other stressors and ecosystem responses.
- **Growing seasons will get longer (virtually certain).** Earlier spring thaws and later first frosts in autumn could result in greater growth and productivity, but only if there is enough water.
- **The nature and timing of precipitation will change (virtually certain).** Annual precipitation may increase, but a greater proportion of precipitation may occur during winter, leaving longer, drier summers.
- **Soil moisture patterns will change (virtually certain), with drier soil conditions later in the growing season (likely).** Changing rainfall patterns and increased evapotranspiration are expected to decrease soil moisture.
- **The frequency, size, and severity of natural disturbances will change across the landscape (very likely).** Wind storms, ice storms, droughts, wildfires, and floods are likely to cause greater damage.
- **Pests and diseases will increase or become more severe (very likely).** Better able to survive warmer winters or complete a second lifecycle in one year, insect pests, nonnative species, and diseases may expand their range and abundance.

Ecosystem Response to Shifting Stressors

Forest ecosystems will continue to adapt to changing conditions.

- **Suitable habitat for many tree species will move northward (virtually certain).** Species at the southern end of their range may experience greater stress as the suitable range moves northward, even as southern and invasive species gain a competitive advantage.
- **Many of the current dominant tree species will decline (likely).** Many species, including balsam fir, white spruce, paper birch, and quaking aspen, are projected to decline as their suitable habitat decreases in quality and extent.

- **Forest succession will change, making future trajectories unclear (very likely).** In response to changes in species distributions, communities may fundamentally change or even disaggregate as increased stress, disturbance, and competition from nonnative species alter competitive dynamics.
- **Interactions of multiple stressors will reduce forest productivity (likely).** Changes in hydrology, disturbances, and other stressors may combine to reduce growth rates, vigor, and health of many important species.

Ecosystem Vulnerabilities

Certain species, communities, and ecosystems may be particularly vulnerable to severe declines in abundance or may be lost entirely from the landscape.

- **Risk will be greater in low-diversity ecosystems (very likely).** Ecosystems dominated by a single species are more likely to experience severe degradation if that species declines.
- **Disturbance will destabilize static ecosystems (very likely).** Systems that are not resilient to disturbances may be particularly vulnerable as natural disturbances increase.
- **Climate change will exacerbate problems for species already in decline (very likely).** Eastern hemlock, northern white-cedar, and yellow birch have been declining in northern Wisconsin. Models project these species' suitable habitat to decrease further.
- **Resilience will be weakened in fragmented ecosystems (very likely).** Smaller, separated patches often support lower species diversity and genetic diversity, reducing the ability of species to adapt or migrate.
- **Altered hydrology will jeopardize lowland forests (very likely).** Altered precipitation regimes could dry peat systems or other sites that rely on saturated soils, leaving them vulnerable to extreme stress or severe wildfire.
- **Changes in habitat will disproportionately affect boreal species (virtually certain).** Projected decreases in potential suitable habitat are especially significant for many boreal species.
- **Further reductions in habitat will impact threatened, endangered, and rare species (virtually certain).** Species with very specific habitat requirements and low resilience will be vulnerable to changes.
- **Ecosystem changes will have significant effects on wildlife (very likely).** Species that rely on trees for food or habitat are likely to be impacted by changes in tree species.

Management Implications

Management practices have always had an important influence on forest composition, structure, and function, and will continue to influence the way that forests respond to climate change.

- **Management will continue to be an important ecosystem driver (virtually certain).** Management practices will continue to shape forests by influencing forest composition, species movement, and successional trajectories.
- **Many current management objectives and practices will face substantial challenges (virtually certain).** Many commercially and economically important tree species may face increased stress and lowered productivity, which may affect availability for some products.
- **More resources will be needed to sustain functioning ecosystems (virtually certain).** Impacts of climate change will increase the human and capital resources needed to assist regeneration of native species, control wildfires, combat invasive species and cope with pests and diseases.

Climate Change Adaptation

Broadly, adaptation includes all adjustments, both planned and unplanned, in natural and human systems in response to climatic changes and subsequent effects (Parry et al. 2007). In this document, however, we focus on planned, ecosystem-based adaptation activities that use a range of opportunities for sustainable management, conservation, and restoration of forests in order to maintain ecosystem integrity and provide environmental benefits to people (Groves et al. 2010, Secretariat of the Convention on Biological Diversity 2009). The discipline of climate change adaptation is rapidly growing (Glick et al. 2011, Heller and Zaveleta 2009), and many broad recommendations for adapting ecosystems to climate change have been suggested and synthesized (e.g., Heinz Center 2008, Heller and Zaveleta 2009, Millar et al. 2007, Ogden and Innes 2008,). Additionally, the growth and development of this topic area over time support the development of overarching principles for incorporating adaptation into management perspectives (Box 2).

Numerous actions can be taken to enhance the ability of ecosystems to adapt to climate change and its effects. These adaptation actions can be used to reduce or avoid loss of forest cover, declines in forest productivity, alterations to ecosystem processes, reductions in the environmental benefits that forests provide to people (such as wildlife, recreation, and wood products), and many other potential impacts on forests. This document provides a set of resources to help forest owners and managers adapt forests to new and changing conditions and sustain healthy ecosystems over the long term. Forests that are well-adapted to climate change and climate variability may be better poised to persist or even thrive under future conditions, as well as to meet goals for forest management.

Box 2: Principles for Adaptation

Land managers have many tools available to begin to address climate change; however, a new perspective will be needed to expand management thinking to consider new issues, spatial scales, timing, and prioritization of efforts (Lawler et al. 2009). The following principles can serve as a starting point for this perspective (Joyce et al. 2008, Millar et al. 2007, WICCI 2011b):

- **Prioritization and triage:** It will be increasingly important to prioritize actions for adaptation based both on the vulnerability of resources and on the likelihood that actions to reduce vulnerability will be effective.

- **Flexible and adaptive management:** Adaptive management provides a decisionmaking framework that maintains flexibility and incorporates new knowledge and experience over time.

- **"No regrets" decisions:** Actions that result in a wide variety of benefits under multiple scenarios and have little or no risk may be initial places to look for near-term implementation.

- **Precautionary actions:** Where vulnerability is high, precautionary actions to reduce risk in the near term, even with existing uncertainty, may be extremely important.

- **Variability and uncertainty:** Climate change is much more than increasing temperatures; increasing climate variability will lead to equal or greater impacts that will need to be addressed.

- **Integrating mitigation:** Many adaptation actions are complementary with actions to mitigate greenhouse gas emissions, and actions to adapt forests to future conditions can help maintain and increase their ability to sequester carbon.

In this document, we build upon the adaptation framework described by Millar et al. (2007). The concepts of resistance, resilience, and response serve as the fundamental options for managers to consider when responding to climate change:

- Resistance actions improve the forest's defenses against anticipated changes or directly defend the forest against disturbance in order to maintain relatively unchanged conditions. Although this option may be effective in the short term, it is likely that resistance options will require greater resources and effort in resisting change over the long term as the climate shifts further from historical norms. Additionally, as the ecosystem persists into an unsuitable climate, the risk that the ecosystem will undergo irreversible change (such as through a severe disturbance) increases over time.

- Resilience actions accommodate some degree of change, but encourage a return to prior conditions after a disturbance, either naturally or through management. Resilience actions may also be best suited to short-term efforts, high-value resources, or areas that are well-buffered from climate change impacts. Like the resistance option, this option may engender an increasing level of risk over time if an ecosystem becomes increasingly ill-suited to the altered climate.

- Response actions intentionally accommodate change and enable ecosystems to adaptively respond to changing and new conditions. A wide range of actions exists under this option, all working to influence the ways in which ecosystems adapt to future conditions, instead of being caught off-guard by rapid and catastrophic changes.

These options of resistance, resilience, and response serve as the broadest and most widely applicable level of a continuum of management responses to climate change (Fig. 2; Janowiak et al. 2011). Along this continuum, actions for adaptation become increasingly specific. Adaptation strategies begin to illustrate the ways that adaptation options could be employed, and are abundant in recent literature and reports. Strategies, however, are still very broad, and can be applied in many ways across a number of landscapes and forest types. Approaches provide greater detail in how managers may be able to respond, and differences in application among specific forest types and management goals start to become evident. Ultimately, tactics are the most specific adaptation response on the continuum, providing prescriptive direction in how actions can be applied on the ground. Throughout this document, we endeavor to provide resources for land managers to consider broad adaptation strategies and approaches, and then work toward devising implementable tactics for creating better-adapted forests.

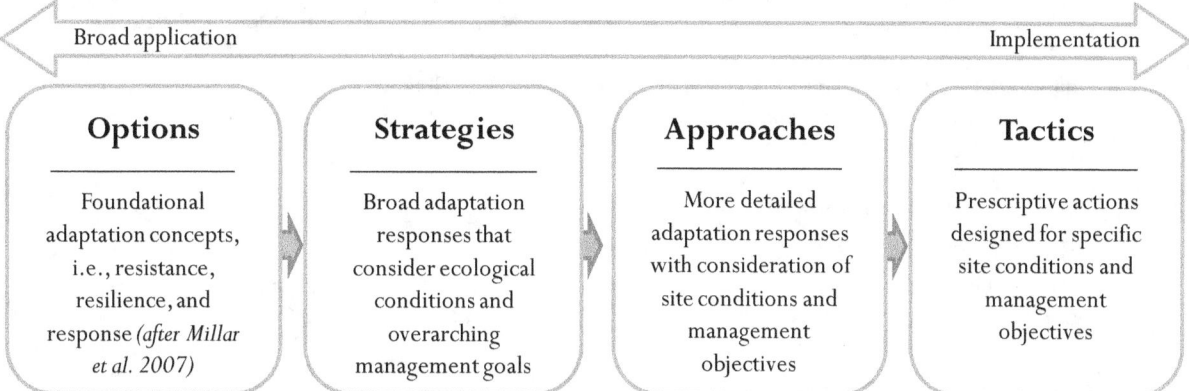

Figure 2.—A continuum of adaptation actions is available to address needs at appropriate scales and levels of management. (Modified from Janowiak et al. 2011.)

Guide to Chapters

Land managers face an immense set of challenges in responding to climate change. Managers must first determine the potential impacts of climate change on a particular location given a wide range of uncertainty about precise future conditions. They then need to incorporate climate change adaptation into existing management plans and policies to meet current management goals, reduce climate change-related risks, and leave options open for future decisions. In this context, it becomes essential that managers begin to use the tools and information that are currently at hand to adjust management goals and actions for long-term ecosystem sustainability.

This document contains a set of interrelated chapters, each of which serves as a resource to help managers incorporate climate change considerations into management and devise adaptation tactics that can be used to respond to climate change. They do not provide recommendations or set policy, but rather provide information, ideas, and processes for evaluating climate change as one component of management (Box 3).

Chapter 1: Framework Overview describes the overarching process developed as part of the Climate Change Response Framework, summarizes the project components and subsequent outcomes, and provides a background for later chapters.

Box 3: About the *Forest Adaptation Resources*

The *Forest Adaptation Resources* can:

- Help managers incorporate climate change considerations into management and devise adaptation actions that can be used to respond to climate change.
- Be used by a variety of land management organizations, including both private and public entities.
- Be approached, read, and used flexibly. Although arranged as a single document, the chapters are designed both to support each other and to be used independently of each other.
- Be applied across multiple scales in different places, through
  - A broad framework for responding to climate change (Chapter 1), currently being applied in the "Northwoods" of Wisconsin, Minnesota, and Michigan, as well as other regions.
  - A "menu" of many adaptation strategies and approaches for forest ecosystems, with additional information and considerations specific to the types of forests present in northern Wisconsin (Chapter 2).
  - An Adaptation Workbook, which can be applied to numerous ecosystems at multiple scales, to help land managers integrate climate change considerations into management activities (Chapter 3).
- Illustrate practical applications of the process through the "illustrations" or case studies of how climate change was considered in real-world management situations using the other resources within this document (Chapter 4).

The resources in this document do not:

- Recommend management actions or policy.
- Replace institutional or legal processes. These resources are not intended to replace existing decisionmaking processes, but may augment them at an organization's discretion.

Chapter 2: Adaptation Strategies and Approaches synthesizes a wide range of reports and peer-reviewed publications on climate change adaptation and provides a "menu" of adaptation actions that are relevant to northern Wisconsin. Expert feedback was used to further refine the existing literature and provide considerations for use of the approaches in 12 different forest types.

Chapter 3: Adaptation Workbook outlines a process for incorporating climate change considerations into management planning and activities, while complementing existing processes and procedures for making decisions. It uses a workbook approach to provide step-by-step instructions for land managers to translate the Adaptation Strategies and Approaches into on-the-ground management tactics that help forest ecosystems adapt to climate change.

Chapter 4: Illustrations demonstrates how the Adaptation Strategies and Approaches and the Adaptation Workbook chapters can be used together to develop adaptation tactics (Fig. 3). Two illustrations of real-world management issues provide examples to managers completing the Adaptation Workbook, and also show how climate change considerations can be incorporated into management planning and decisionmaking.

Adaptation Strategies and Approaches:
Presents a "menu" of adaptation strategies and approaches for northern Wisconsin forests

Adaptation Workbook:
Outlines a series of steps for incorporating climate change into existing management

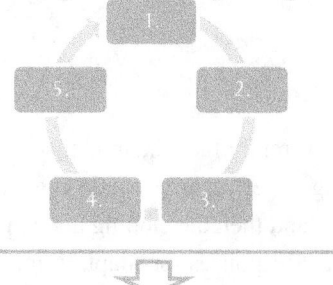

Illustrations:
Provides real-world examples of how the above are used together to develop tactics for adaptation

Figure 3.—The Adaptation Strategies and Approaches and Adaptation Workbook chapters can be used together to develop management tactics to adapt forests to the anticipated effects of climate change.

CHAPTER 1: CLIMATE CHANGE RESPONSE FRAMEWORK OVERVIEW

Chris Swanston, Maria Janowiak, and Patricia Butler

Managers currently face the immense challenge of anticipating the effects of climate change on forest ecosystems and then developing and applying management responses for adapting forests to future conditions. The Climate Change Response Framework (CCRF) is a highly collaborative approach to helping land managers understand the potential effects of climate change on forest ecosystems and integrating climate change considerations into management. Since 2009, the CCRF project in northern Wisconsin has worked to bridge the gap between scientific research on climate change impacts and on-the-ground management. Climate change has the potential to affect areas much larger than any single ownership, making multi-institutional efforts and partnerships crucial in addressing climate change. The information and tools developed as part of this project can be applied by forest owners and managers working in a variety of private and public agencies, both within and beyond northern Wisconsin. Currently, the framework is being applied in other locations in the eastern United States through coordinated place-based projects (Northern Institute of Applied Climate Science [NIACS] 2011). In this chapter, we outline this general framework for adapting to climate change. We then describe the application of the framework through project activities in northern Wisconsin and provide background for the subsequent sections of this document.

A Framework for Adapting to Climate Change

From its beginning, the CCRF was conceived as a model for collaborative management and climate change response across large and diverse landscapes that could be employed beyond the original geographic borders of the project (NIACS 2011). With this in mind, the framework represents a broad approach for responding to climate change (Fig. 4) that can be adjusted and applied to other locations and landscapes. The overall process is adaptive and incorporates opportunities for new information, ideas, and lessons learned during the process to be incorporated into the elements and activities.

Identify location, ecosystems, and time frame

The first step in this process is defining the scope of the project: the geographic scale and extent of the analysis area, the ecosystems of interest, and relevant timelines for evaluating available information. The geographic scale at which the project works is very important and needs to be chosen so that an adequate level of detail is available in the assessment stage for later use in management decisions. If the analysis area is too large, the information may be too general and unsuitable for supporting decisions in specific management applications. If the analysis area is too small, however, available information

1. Identify location, ecosystems, and time frame.

2. Establish partnerships.

3. Assess ecosystem vulnerabilities and mitigation potential.

4. Compile adaptation strategies and approaches.

5. Plan and implement at appropriate scales.

6. Integrate monitoring and evaluate effectiveness.

Figure 4.—The Climate Change Response Framework uses an adaptive management approach to help land managers understand the potential effects of climate change on forest ecosystems and integrate climate change considerations into management.

(e.g., climate projections, impact model results) developed at much larger spatial scales may not be suitable for use at the finer spatial resolution of the area. The location, scale, ecosystems, and time frame identified can be refined throughout the process as needed. For example, the time frame may be influenced by the information that is selected for use in the assessments.

Establish partnerships

Climate change is inherently a cross-boundary issue because all places will be affected in some way, regardless of ownership. Communication and coordination with partners increase the ability of everyone involved to respond to climate change by increasing the amount and accessibility of information and ideas. It is critical to bring key partners into the process as early as possible. Whenever possible, the use of existing partnerships is extremely valuable because it builds upon established relationships and can avoid reinventing the wheel in many situations. However, it is also

very important to look for new partners. Given the broad impacts and importance of climate change, new partners may be available beyond those that are normally enlisted. This step is critical in defining and launching a new project, but never really stops thereafter. A successful effort will draw new partners, who will continue to enrich the skills, creativity, lands, resources, and perspectives available to the project. Profiting from partnerships requires flexibility, continual focus on project objectives and timelines, and excellent communication.

Assess ecosystem vulnerabilities and mitigation potential

Collecting information about projected changes in climate and impacts on ecosystems provides critical information to determine what species, ecosystems, or other features are most vulnerable to the effects of climate change. Likewise, assessing the current amount of carbon and the mitigation potential of the analysis area can provide information for consideration in management decisionmaking. There is no universally accepted process for developing these assessments, and new and different assessments are rapidly being developed (e.g., Byers and Norris 2011, Doppelt et al. 2009, Swanston et al. 2011). *Scanning the Conservation Horizon: A Guide to Climate Change Vulnerability Assessment* (Glick et al. 2011) outlines the elements of a vulnerability assessment and provides examples of possible approaches for creating an assessment. In general, assessments should strive to include both a synthesis of existing information and new information on the ecosystems of interest. Drawing upon partners, especially research partners, is a great way to develop new information. Additionally, a "common-sense filter" should be used to put analyses in perspective. Scientific models or other information used to create new information may not provide a complete picture of on-the-ground realities; input from both scientists and managers can be important in putting scientific results in context.

Compile adaptation strategies and approaches

A wealth of information is available on the adaptation of forests to the effects of climate change, as well as on the potential for forests to mitigate greenhouse gases. However, most of this information is very broad and not directly applicable at scales most relevant to land managers. Re-examining the breadth of adaptation strategies being discussed in scientific and management communities within the specific context of the analysis area and its ecosystems will help identify an array of strategies that are the most relevant for local land managers. This comprehensive array of strategies and approaches does not focus on a particular land use or management goal, but instead serves as a "menu" from which managers can select actions based on their management needs for a particular situation. In fact, some strategies may be mutually exclusive in a given place, but can be applied simultaneously in different places across the landscape according to management goals and ecological considerations. Integrating local vulnerabilities with a menu of adaptation approaches can help managers devise the most realistic adaptation tactics for their needs.

Plan and implement at appropriate scales

After considering management goals and local vulnerabilities, and then choosing adaptation strategies and approaches, land owners and managers can devise adaptation tactics that are best suited to their needs and constraints. Just as there is a menu of adaptation strategies and approaches, additional tactics may be created to increase the range of choices under consideration. The implementation of some tactics may be considered practicable and appropriate in some ownerships, but not in others. Even within an ownership, some tactics may be deemed fully practicable and even necessary in the long term, but too risky or uncertain in the near term. Implementation of adaptation tactics will vary widely across ownerships and through time in the same way that a wide variety of tactics is currently applied in forest management.

Integrate monitoring and evaluate effectiveness

Monitoring is a critical step to evaluate whether management actions are effective in responding to climate change and reducing the vulnerability of ecosystems to changes that are occurring. As with implementation, the ways in which monitoring is implemented and monitoring results are evaluated and incorporated into management will depend upon land managers' particular decisionmaking processes and plans. Results from monitoring can be integrated throughout this framework to refine individual steps. For example, monitoring results may be able to provide more detail on the vulnerability of ecosystem components to specific climate change impacts, and this information could be included in relevant assessments.

Applying the Framework in Northern Wisconsin

The CCRF was developed to synthesize information on climate change to help forest owners and managers incorporate climate change considerations into land management planning and activities. The CCRF project underway in northern Wisconsin is the result of a substantial collaboration among multiple organizations, with the Chequamegon-Nicolet National Forest (CNNF) playing a critical role as an initial test-bed for this project. In this section, we describe the application of the framework in northern Wisconsin, including activities and products developed as components of the project.

Early in the process, we identified flexibility and communication as critical features in the project. Allowing for a certain amount of flexibility gave us the opportunity to make needed adjustments and capitalize on opportunities as they arose. We developed robust and productive communication avenues between the major partners and communicated frequently with potential users. Ultimately, however, we identified the need for much stronger and earlier outreach to potential users of the products.

Identify location, ecosystems, and time frame

Although the CCRF was catalyzed by the Forest Service's interest in gaining information and insights about how to respond to climate change, we determined at the outset of the project that it would be critical to work beyond National Forest boundaries. Climate change will affect forests regardless of ownership boundaries, and information and tools designed to support informed responses to climate change within the CNNF are relevant to all land managers in northern Wisconsin.

We defined this project's area of interest, the "analysis area," as the portion of Wisconsin that is within Ecological Province 212 (Fig. 5; ECOMAP 1993). This area of Mixed Laurentian Forest, generally referred to as northern Wisconsin in this document, has ecological, political, and social significance. A time frame of approximately 100 years for the assessments was identified based on the majority of information that is available on projected climate change impacts. Additional time frames may be evaluated in subsequent versions of the assessment.

The ecological emphasis of the project is forested ecosystems. Forests are the primary land use in northern Wisconsin, covering 46 percent of the area (Wisconsin Department of Natural Resources 1998). Other ecosystems and methods of assessment may be added to the project in the future. For example, a watershed vulnerability assessment project that is

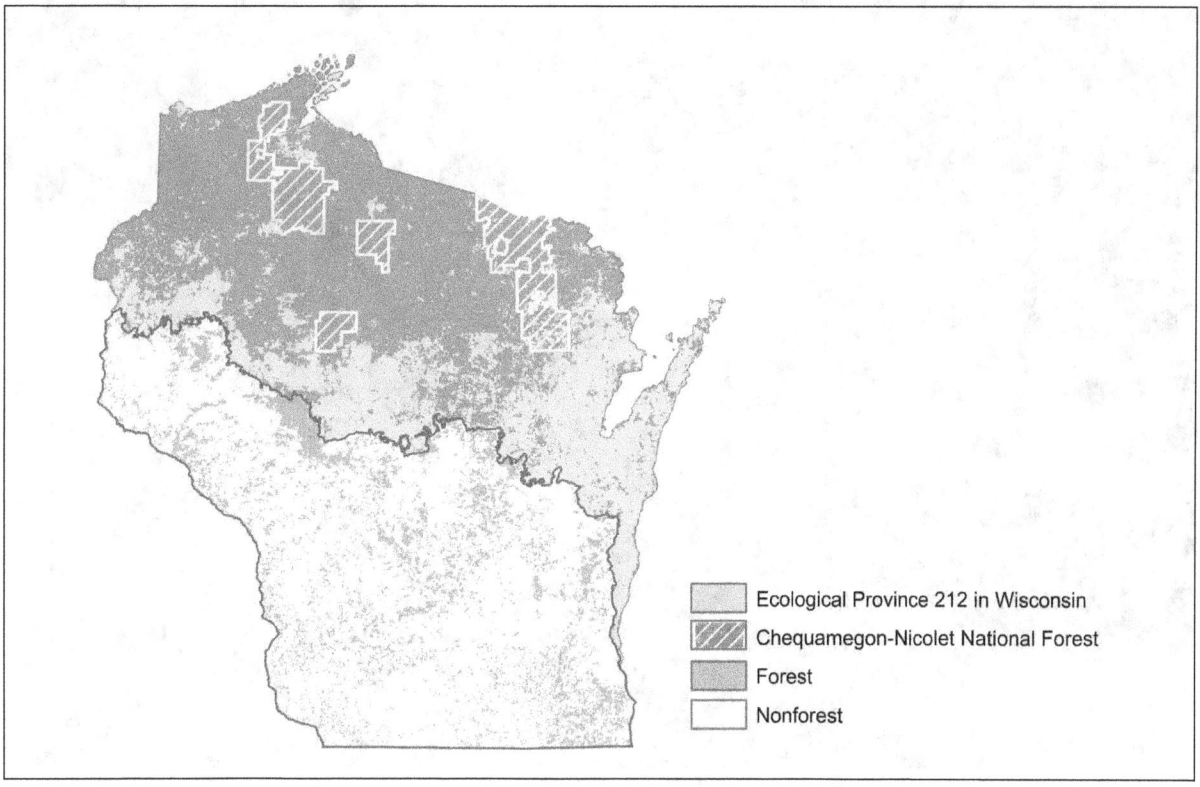

Figure 5.—The analysis area is the portion of Ecological Province 212 (Mixed Laurentian Forest) within Wisconsin.

occurring at the CNNF may help to expand the scope of future versions of the vulnerability assessment.

Establish partnerships

We sought to bring in a variety of partners early in the project. This included the Forest Service Northeastern Area, State and Private Forestry, which strongly encouraged expanding partnerships and was instrumental in bringing in additional landowners and stakeholders. The Wisconsin Department of Natural Resources and the Wisconsin Initiative on Climate Change Impacts (WICCI) were also engaged as key partners very early. Coordination between our project and the Forestry Working Group of WICCI also allowed each group to focus on different regions of the state in pursuit of similar goals.

The Shared Landscapes Initiative was a fundamental component for establishing communication and partnerships across a broad spectrum of forest land owners, managers, and the general public. The purpose of this effort is to foster dialogue about climate change, ecosystem response, ecosystem management, and cooperative activities among these different groups. Working with the Northeastern Area and the CNNF, we hosted a Shared Landscapes Initiative workshop in northern Wisconsin in February 2010. More than 70 people attended the workshop, and the Initiative currently includes more than 30 organizations and individuals that represent federal, state, other public, tribal, and private interests. A Shared Landscapes Work Group was established to maintain communication within the Initiative and facilitate creation of cooperative projects and products.

Trees surrounding the edge of a small wetland in northern Wisconsin.

Photo by Maria K. Janowiak, U.S. Forest Service and Northern Institute of Applied Climate Science

We also expanded partnerships within the scientific community by hosting a Climate Change Science Applications and Needs workshop in April 2010. This workshop brought together more than 50 participants, including scientists from northern Wisconsin, the Great Lakes, and elsewhere in the United States. Participants discussed science needs, applications of science, and climate change monitoring in northern Wisconsin. Workshop participants were invited to join a Climate Change Science Roundtable, whose members have contributed to the scientific review of project documents. As the project itself continues to evolve, the Roundtable is expanding to provide climate-related scientific information and perspective to the Shared Landscapes Initiative. Many of the ideas and conversations from this workshop, Roundtable discussions, and follow-up engagement with workshop participants were incorporated into other aspects of this project, including the assessments and this document.

Assess ecosystem vulnerabilities and mitigation potential

We developed the Ecosystem Vulnerability Assessment and Synthesis for northern Wisconsin (Swanston et al. 2011). This assessment describes the contemporary landscape of northern Wisconsin, summarizes projected changes in northern Wisconsin climate at the end of the century, presents results from two vegetation impact models used to project changes in forest composition across a range of potential climates, and assesses the implications of climate change for forest ecosystems in northern Wisconsin. The assessment drew upon and expanded existing scientific examinations of climate change in northern Wisconsin. Downscaled climate projections were readily available from WICCI, and were fundamental to the assessment document and process. Impact models are useful for projecting impacts of climate change on species and ecosystems, but operate under certain assumptions and limitations that contribute to

inherent uncertainty (Swanston et al. 2011, Wiens et al. 2009). For this reason, we used two very different impact models—LANDIS-II (a process model) and the Climate Change Atlas (a species distribution model)—to examine changes to forest ecosystems, and then worked with the results of both models to identify species and ecosystems with greater certainty of future impacts.

A mitigation assessment in preparation will provide information on current carbon stocks in northern Wisconsin and summarize available information on how forest management, land use, and other changes could alter the amount of carbon stored in forests and wood products. It will also contain more detailed results on how carbon stocks may be altered in the future as a result of climate change and forest management. As with the vulnerability assessment, the mitigation assessment will be made publically available in a published format.

Compile adaptation strategies and approaches

We created a set of Adaptation Strategies and Approaches for forests in northern Wisconsin (Chapter 2). These approaches and strategies were synthesized from a broad set of literature and other resources, and refined using feedback from scientists and managers with expertise in regional forest types. Although several documents describing strategies for adaptation have recently been published (e.g., Galatowitsch et al. 2009, Heinz Center 2008, Millar et al. 2007, National Research Council 2010), the vast majority of the adaptation strategies that have been described are general in nature and apply only broadly to many ecosystems and forest types. The central goal in developing our set of strategies and approaches was to focus on a particular location (i.e., northern Wisconsin) and ecosystem (i.e., forests) to provide a relevant and useful list of actions that can be pursued in regional forests. Throughout the development of these actions, we asked ourselves a series of questions to maintain our focus: Within northern Wisconsin, what are the anticipated effects of climate change? What is an approach for addressing these impacts and helping forests to adapt? How might the use of this approach vary for different forest types?

Plan and implement at appropriate scales

Recognizing that forest owners and managers would ultimately be making decisions on which actions to implement to respond to climate change, we concentrated on developing information and resources that could be used to support these decisions across a variety of land ownerships. We have worked to avoid making recommendations or interfering with existing management planning and decisionmaking; rather, products of the CCRF, including this document, are designed to supplement existing procedures.

Several components of this project can be used to support management decisions for responding to climate change. *The Ecosystem Vulnerability Assessment and Synthesis* (Swanston et al. 2011) provides an extensive description of potential climate change impacts on forests, and additional information that becomes available will be presented in future versions of the assessment. Further, three interrelated chapters of this document help incorporate climate change considerations into management decisions and lead managers toward devising management tactics that can be used to respond to climate change. The Adaptation Workbook (Chapter 3) sets out a process for managers to assess the potential climate change impacts for a management project or issue and then to evaluate the Adaptation Strategies and Approaches that could be used based on their specific management objectives, site conditions, and other factors. Two Illustrations (Chapter 4) provide

examples of how the Adaptation Workbook can be used based on our work with two teams from the CNNF to test and refine the Adaptation Workbook.

Integrate monitoring and evaluate effectiveness

Similar to implementation, necessary and appropriate monitoring will vary by forest owner, site conditions, management objectives, management activities, and relevant time frames. We included monitoring as a vital step within the Adaptation Workbook so that managers will identify what monitoring is necessary to determine whether implemented adaptation actions effectively meet management objectives and adapt forests to changing conditions. It will be beneficial and efficient for monitoring activities to build upon existing efforts when possible (see Appendix 1 for a list of monitoring activities in northern Wisconsin).

Summary

Through the CCRF, we have outlined a general approach for responding to climate change (Fig. 4) that can be adjusted and applied to other locations and landscapes. We have applied this process in northern Wisconsin, and will continue to work with landowners across northern Wisconsin to pursue this process and help forests adapt to the impacts of climate change. At the same time, we have begun extending this project to include all of the Northwoods of Michigan, Wisconsin, and Minnesota, as well as other regions. We recognize that this expanded effort will be able to draw greatly from the work done in Wisconsin, but it will also be different in many ways as we incorporate new partners, work across a larger geographic area, and collaborate with other groups working within different environmental and social systems.

CHAPTER 2: ADAPTATION STRATEGIES AND APPROACHES

Patricia Butler, Chris Swanston, Maria Janowiak, Linda Parker,
Matt St. Pierre, and Leslie Brandt

A wealth of information is available on climate change adaptation, but much of it is very broad and of limited use at the finer spatial scales most relevant to land managers. This chapter contains a "menu" of adaptation actions and provides land managers in northern Wisconsin with a range of options to help forest ecosystems adapt to climate change impacts (Box 4). Land managers may select strategies and approaches from this menu based on their management goals and needs. This chapter also provides a basis for the Adaptation Workbook (Chapter 3), where managers consider local site conditions, climate change vulnerabilities, and

other factors to further refine adaptation approaches into specific tactics that can be implemented on the ground.

About the Adaptation Strategies and Approaches

Information on how to adapt natural ecosystems to the anticipated effects of climate change is rapidly growing as increasing numbers of land managers engage with the topic (e.g., Gunn et al. 2009, Heinz Center 2008, The Nature Conservancy [TNC]

Box 4: Using the Adaptation Strategies and Approaches

The Adaptation Strategies and Approaches can:

- Present a full spectrum of possible adaptation actions that can help sustain forests and achieve management goals in the face of climate change.
- Serve as a "menu" of adaptation actions—managers select actions best suited to their specific management goals and objectives.
- Provide co-workers, team members, and other collaborators with a platform for discussion of climate change-related topics and issues.

The Adaptation Strategies and Approaches do not:

- Make recommendations or set guidelines for management decisions.
- Express preference for use of any of the strategies and approaches within a forest type, location, or situation. Rather, a combination of location-specific factors and manager expertise is meant to inform the selection of any strategy or approach.

2009). Much of this information, however, remains broadly applicable across ecosystem types without regard to geographic setting and does not provide sufficient detail for forest managers to identify specific response actions. We have compiled and synthesized a list of regionally focused strategies and approaches that may be used to adapt the forests of northern Wisconsin to a range of anticipated climate conditions and ecosystem impacts.

Importantly, the adaptation strategies and approaches presented in this chapter are nested within the existing paradigm of sustainable forest management. A changing climate and the associated uncertainty will create many challenges, forcing managers to be flexible and make adjustments to management objectives and techniques; however, the overarching goal of sustaining forests over the long term will remain a cornerstone of management. Many actions to adapt forests to climate change are consistent with

sustainable management (Innes et al. 2009, Ogden and Innes 2008) and efforts to restore ecosystem function and integrity (Harris et al. 2006, Millar et al. 2006). Additionally, many current management activities make positive contributions toward increasing forest health and resilience in the face of climate change.

The strategies and approaches in this document are part of a continuum of adaptation actions ranging from broad, conceptual application to practical implementation (Fig. 6). They were synthesized from many scientific papers that discussed adaptation actions at numerous scales and locations (Appendix 2). Feedback from experts, including regional scientists, forest ecologists, and others, was used to refine the strategies and approaches and to provide information about the potential use of approaches for individual forest types in northern Wisconsin (Table 1).

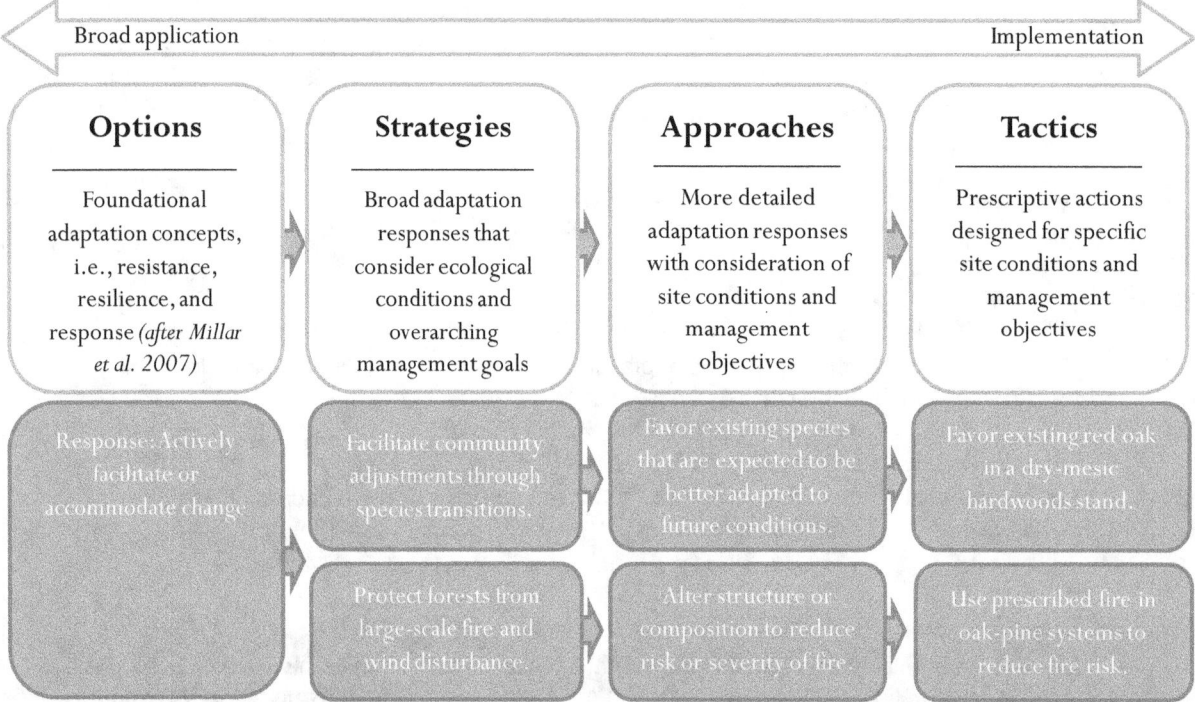

Figure 6.—A continuum of adaptation actions is available to address needs at appropriate scales and levels of management (top). The shaded boxes provide examples of each level of action (bottom). (Modified from Janowiak et al. 2011.)

Table 1.—Descriptions of the forest types present in northern Wisconsin. Common and scientific names for species are presented in Appendix 3.

Aspen – Dominated by quaking aspen, bigtooth aspen, or balsam poplar. Some stands may have co-dominant tree species such as balsam fir or white spruce.

Balsam Fir – Dominated by balsam fir. Some stands may include a component of quaking aspen or paper birch.

Hemlock – Dominated by eastern hemlock. Yellow birch and sugar maple are sometimes co-dominant.

Jack Pine – Stands are generally dominated by jack pine, with some stands being composed primarily of mixed pine species or on a rare occasion Scotch pine. Oak species may be co-dominant in some stands.

Lowland Conifer – Stands in lowland sites that are dominated primarily by black spruce, northern white-cedar, tamarack, or a mixture of these conifer species. Quaking aspen, paper birch and other species may be co-dominant in some stands.

Lowland Hardwood – Stands in lowland sites that are dominated primarily by black ash, red maple, or American elm, or a mixture of these species.

Northern Hardwood – Stands composed largely of sugar and red maple. Hemlock, yellow birch, basswood, red oak, and black cherry are also likely to be found in varying amounts, depending on site conditions.

Oak – Dominated by one or more oak species. Aspen, white pine, and other species may be co-dominant in some stands.

Paper Birch – Dominated by paper birch. Some stands may contain components of aspen or balsam fir.

Red Pine – Dominated by red pine. Some stands have an oak component in the understory and sometimes oak is a co-dominant.

Spruce – Generally dominated by white spruce (occasionally black spruce or Norway spruce). Some white spruce stands may have co-dominant tree species, such as balsam fir or quaking aspen.

White Pine – Dominated by eastern white pine. Some stands may include a component of hemlock or northern red oak and white ash.

At a broad level, the strategies presented below vary in their emphasis on resisting potential changes, building resilience, or actively responding to accommodate change (Table 2). The approaches nested under each strategy further vary in management intensity and style, emphasis on form or function, and reliance on traditional or experimental techniques. Some are more oriented toward passively adjusting to forest changes, while others seek to more actively guide changes. The challenge is to select the approaches that not only meet existing management goals, but also help prepare forests for a range of potential future climates and conditions.

The level of detail provided in this chapter helps tailor the approaches to the ecosystems of this region. A single adaptation approach may be applicable for many forest types but be executed in very different ways in each individual forest type. However, it is critical to keep in mind that site conditions, forest types, and management history and goals will all influence whether an approach is appropriate for a particular situation. Not all approaches can or should be used in every location or situation; rather, when a range of adaptation options (i.e., resistance, resilience, and response) exist, some of the approaches may be mutually exclusive or generally counteract each other.

We have added text to many of the approach descriptions that highlight examples or special considerations within a given forest type to help address the different ways in which an approach may be applied. We provide examples to describe how an approach might be implemented in order to spur the generation of other examples or applications. We also provide considerations for using the approach in specific forest types. These considerations are not exhaustive, and there are likely to be many more considerations unique to a forest type than are provided. Similarly, even if a forest type is not specifically highlighted, the approach may still be considered for use.

Table 2.—Climate change adaptation strategies under three broad adaptation options.

Strategy	Resistance	Resilience	Response
1. Sustain fundamental ecological functions.	X	X	X
2. Reduce the impact of existing biological stressors.	X	X	X
3. Protect forests from severe fire and wind disturbance.	X	X	
4. Maintain or create refugia.	X		
5. Maintain and enhance species and structural diversity.	X	X	
6. Increase ecosystem redundancy across the landscape.		X	X
7. Promote landscape connectivity.		X	X
8. Enhance genetic diversity.		X	X
9. Facilitate community adjustments through species transitions.			X
10. Plan for and respond to disturbance.			X

Adaptation Strategies and Approaches

Strategy 1: Sustain fundamental ecological functions

Climate change will have substantial effects on a suite of ecosystem functions, and many management actions will need to work both directly and indirectly to maintain the integrity of ecosystems in the face of climate change. This strategy seeks to sustain fundamental ecological functions, including those related to soil and hydrologic conditions. Adaptation approaches under this strategy should be used in concert with other approaches to maintain ecosystem productivity and health, as well as to meet management goals and objectives.

Approaches:

Maintain or restore soil quality and nutrient cycling

Northern Wisconsin is experiencing increased temperatures (Kucharik et al. 2010), which correspond to shorter periods of snow cover and frozen water and soils. Maintaining both soil quality and nutrient cycling in forest ecosystems is already a common tenet of sustainable forest management

(e.g., Wisconsin Department of Natural Resources [WDNR] 1995), and continued vigilance will help improve the capacity of the forest to persist under new conditions. Re-evaluation of the timing and intensity of some practices will help ensure that site quality is not degraded as both ecosystem vulnerabilities and the duration of seasons change. One example of an adaptation tactic under this approach is to alter the timing of logging operations to prevent soil compaction, realizing that the time when soils will be frozen or protected by snowpack is decreasing (Gunn et al. 2009). Another example of a tactic is to retain coarse woody debris in order to maintain moisture conditions, soil quality, and nutrient cycling (Covington 1981, Duvall and Grigal 1999).

Additional considerations for individual forest types under this approach include:

- **Aspen:** This forest type is expected to be sensitive to increased temperature and decreased precipitation (Burns and Honkala 1990). Aspen on very dry sites may undergo reduced productivity or vigor as growing seasons continue to lengthen, possibly combined with increased late-season drought.

- **Lowland Conifer:** Very moist soils are critical for this forest type, making it especially susceptible to soil compaction and rutting. Actions to modify timing of entry, equipment, or technique may help to minimize these impacts where optimal windows for winter harvest (e.g., under frozen soil conditions) are shorter as a result of climate change.
- **Spruce:** White spruce tends to be shallow rooted and prefers moist sites. Actions to modify timing of entry, equipment, or technique may help to minimize these impacts where optimal windows for winter harvest (e.g., when snowpack is 6 inches or more) are shorter as a result of climate change.

Maintain or restore hydrology

Some forest types, such as lowland hardwoods and lowland conifers, are very susceptible to drought and may become more vulnerable as a result of climate change. Conversely, other forest types are susceptible to flooding and ponding, which may occur more often as a result of more frequent severe weather events. In order to maintain appropriate hydrologic regimes within systems, existing infrastructure that diverts water or otherwise alters hydrology can be reevaluated to compensate for changes in water levels or flows. It is important to keep in mind that modifications that maintain hydrology at one site may negatively impact hydrology at another site. Examples of adaptation tactics under this approach include minimizing road networks, adjusting culvert size requirements for changes in peak flow, and planning for seasonal limitations on heavy equipment.

Additional considerations for individual forest types under this approach include:

- **Lowland Conifer, Lowland Hardwood:** Maintaining sufficient water level and movement are critical to forest productivity in these forest types and influence species composition. Actions to modify road networks, culverts, or other control points to account for changes in water flow resulting from climate change may help to maintain suitable hydrologic conditions.

Maintain or restore riparian areas

Riparian forests help to buffer stream temperatures as well as increase landscape connectivity for the migration of species (Heller and Zavaleta 2009). The use of best management practices and riparian management zones can be used to avoid damage to riparian areas during management activities. An example of an adaptation tactic under this approach would be to promote conifer species in order to maintain cooler stream temperatures and stream shading. Another example focused at a landscape level could include the reforestation of riparian areas in agricultural areas to reduce erosion into adjacent water bodies.

Additional considerations for individual forest types under this approach include:

- **Balsam Fir, Hemlock, Lowland Conifer:** These forest types often occur in or adjacent to riparian areas. Many conifer species are projected to decline as a result of climate change, and riparian areas may serve as natural refugia for these species.
- **Lowland Hardwood:** Longer growing seasons and drier conditions late in the growing season may result in lower water tables and increased stress on riparian ecosystems. A decline in riparian tree cover may result in a greater increase in stream temperatures and increase the risk of erosion. Actions to manage water levels, such as through the manipulation of existing dams and water control structures, may help to supply proper soil moisture to forests adjacent to the stream during critical time periods.

Strategy 2: Reduce the impact of existing biological stressors

Biological stressors such as insects, pathogens, invasive species, and herbivores can act individually and in concert to amplify the effects of climate change on ecosystems. Forest managers already work to maintain the ability of forests to resist stressors; as an adaptation strategy, these efforts receive added focus and resources, with an emphasis on anticipating and preventing increased stress before it occurs. Dealing with these existing stressors is one of the most valuable and least risky strategies available for climate change adaptation, in part because of the large existing body of knowledge about their impacts and solutions (Climate Change Wildlife Action Plan Work Group 2009).

Approaches:
Maintain or improve the ability of forests to resist pests and pathogens
Even modest changes in climate may cause substantial increases in the distribution and abundance of many forest insects and pathogens (Ayres and Lombardero 2000, Dukes et al. 2009). These impacts may be exacerbated where site conditions, climate, other stressors, and interactions among these factors increase the vulnerability of forests to these agents. Actions to manipulate the density, structure, or species composition of forests may reduce the susceptibility of forests to some pests and pathogens. One example of an adaptation tactic under this approach is to discourage infestation of certain insect pests by reducing the density of a host species and increasing the vigor of the remaining trees. Another example is to maintain an appropriate rotation length to decrease the period of time that a stand is vulnerable to insect pests and pathogens (Spittlehouse and Stewart 2003), recognizing that species are uniquely susceptible to pests and pathogens at various ages and stocking levels. Existing management tactics can also be used to reduce the susceptibility of forests to insects and diseases that may be exacerbated by climate change (Coakley et al. 1999).

Tamarack trees on the edge of a small peatland.

Photo by Maria K. Janowiak, U.S. Forest Service and Northern Institute of Applied Climate Science

Prevent the introduction and establishment of invasive plant species and remove existing invasives

A number of invasive plant species are currently a threat to the forests of northern Wisconsin (WDNR 2009), and climate change is expected to increase invasive species' rate of invasion and spread (Millar et al. 2007). Current recommendations for controlling invasives in forests emphasize early detection of and rapid response to new infestations. An example of a tactic already in practice in northern Wisconsin is the use of guidelines to prevent the spread of invasives by equipment during site preparation or harvesting (WDNR 2009).

Manage herbivory to protect or promote regeneration

Climatic changes may increase the potential for herbivory if populations are able to increase under warmer conditions (Dale et al. 2001, Wisconsin Initiative on Climate Change Impacts [WICCI] 2011b). While many herbivores are present in northern Wisconsin, including insects, rodents, and ungulates, much of the information that is currently available focuses on white-tailed deer (e.g., Waller and Alverson 1997, Waller 2007). Deer herbivory is currently a stressor in some northern Wisconsin forests (Mladenoff and Stearns 1993, Rooney and Waller 2003, WDNR 2010), and there is potential for herbivory to increase in extent and intensity if projected changes in climate lower winter mortality and allow deer populations to grow (WICCI 2011b). As climate change exacerbates many forest stressors, it will be increasingly important to protect regeneration of desired species from deer and other herbivores. An example of an existing tactic that is sometimes employed to influence landscape-level deer use is to perform timber harvests in upland forests to reduce deer migration toward adjacent lowland conifer forests, where regeneration is highly vulnerable to browse. Examples of adaptation tactics at the stand level include the use of fencing and other barriers, as well as "hiding" more desirable species in a mixture of less palatable ones.

Additional considerations for individual forest types under this approach include:

- **Hemlock, Lowland Conifer, Red Pine, White Pine:** Conifer species within these forest types are expected to experience declines in suitable habitat as a result of climate change. Actions to reduce deer browse where it is likely to hinder regeneration of desired species may be especially valuable for helping these species persist under changed conditions.

Strategy 3: Protect forests from severe fire and wind disturbance

Climate change is projected to increase the frequency and severity of droughts and severe weather (Intergovernmental Panel on Climate Change 2007, WICCI 2011b), potentially leading to increased risk of many disturbances, including fire, wind storms, and ice storms (Dale et al. 2001). These disturbances can alter forests over large landscapes and strongly interact with many other stressors. Even as trends continue to emerge, management will need to adjust appropriately to the changes in natural disturbance dynamics (Heller and Zavaleta 2009).

Approaches:

Alter forest structure or composition to reduce risk or severity of fire

Current forest structure and composition may interact with longer and drier growing seasons to increase the risk of fire and associated disturbances (e.g., insect and pathogen outbreaks leading to tree mortality and increased fire risk). Forest management actions to alter species composition or stand structure may increase stand vigor and reduce susceptibility to these threats. An example of an adaptation tactic under this approach is the use of prescribed burning or other ground cover management to minimize fuel loading and reduce the severity of potential fires (Ogden and Innes 2008, Spittlehouse and Stewart 2003). Another example is to plant fire-resistant species, such as hardwoods, between more flammable conifers to reduce vulnerability to wildfires (Spittlehouse and Stewart 2003); this tactic would need to be implemented now to be effective several decades in the future.

Establish fuelbreaks to slow the spread of catastrophic fire

Projected increases in fire as a result of climate change are expected to increase demand on fire-fighting resources and may force prioritization of fire suppression efforts to targeted areas (Millar et al. 2007, Spittlehouse and Stewart 2003). A fuelbreak is defined as a physical barrier to the spread of fire, such as a road, bulldozer line, or body of water; it can also be defined as a change in composition and density of a forest at its edges to reduce fuels. Fuelbreaks can be created to lessen fire spread and intensity in specific areas, such as the wildland-urban interface. Where this approach is designed to protect areas of high value or high concern, the potential for increased forest fragmentation may also be a consideration. An example of an adaptation tactic under this approach is to create a fuelbreak between a flammable or fire-adapted stand and a stand where fire would be undesirable; for example, planting maple between lowland conifer forests and upland fire-prone oak forests may prevent surface fires from moving through the moisture-rich maple leaf litter (Agee et al. 2000). An example of a tactic that is already in practice in northern Wisconsin is to reduce the density of balsam fir (a ladder fuel) in wildland-urban interface areas or near power lines to reduce the spread of fire.

Additional considerations for individual forest types under this approach include:

- **Jack Pine:** Wildfire in jack pine can be difficult to control (Carey 1993). Actions to reduce the risk of large and severe fire, such as the creation of physical fire breaks, in this forest type may help protect refugia or other areas.

- **Lowland Conifer:** Changes in temperature, precipitation, and hydrology may make this system more susceptible to wildfire. The establishment of physical fuelbreaks that displace wetland soils, such as bulldozer lines, has the potential to negatively affect soil structure or hydrology if actions are not taken to minimize impacts.

Alter forest structure to reduce severity or extent of wind and ice damage

Wind disturbances are a fundamental process in many forest ecosystems of the Great Lakes region (Frelich 2002). Wind events and the ensuing effects on forests are expected to become more frequent and severe under climate change (Fischlin et al. 2009, Frelich and Reich 2010), although there are many challenges in predicting the size, frequency, and intensity of these events (Peterson 2000). Some stands may have structures poorly suited to withstand projected increases in storm intensity. Silvicultural techniques exist to alter forest composition and structure for increased resistance to blow-down during severe wind events, although fewer techniques may be available to reduce potential for ice damage. An example of a tactic that is already in practice in northern Wisconsin is to retain trees at the edge of a clearcut to help protect trees in the adjacent stand that have not been previously exposed to wind.

Strategy 4: Maintain or create refugia

Refugia are areas that have resisted ecological changes occurring elsewhere, often providing suitable habitat for relict populations of species that were previously more widespread (Millar et al. 2007). For example, during previous periods of rapid climate change, at-risk populations persisted in refugia that avoided extreme impacts on climate (MacDonald et al. 1998, Millar et al. 2007, Noss 2001). Refugia enable long-term retention of plants, which can then be used to augment the establishment of new forests. This strategy seeks to identify and maintain ecosystems that: (1) are on sites that may be better buffered against climate change and short-term disturbances, and (2) contain communities and species that are at risk across the greater landscape (Millar et al. 2007, Noss 2001). Refugia may or may not function as reserve areas; management activities may be needed to create or maintain refugia.

Approaches:

Prioritize and protect existing populations on unique sites

Some northern Wisconsin ecosystems, such as lowland forests and ephemeral ponds, may be more vulnerable to the impacts of climate change due to their dependence on a narrow range of site conditions. Soil characteristics, hydrologic conditions, topographic variation, and other characteristics may provide conditions that retain habitat for native species and resist invasive species. Existing ecosystems may be more easily maintained at sites with these unique conditions. An example of an adaptation tactic under this approach that focuses on prioritization is to identify unique sites that are expected to be more resistant to change, such as spring-fed stands sheltered in swales, and emphasize maintenance of site quality and existing communities. A more active adaptation tactic is to identify a suite of potential sites for refugia and commit additional resources to ensuring that the characteristic conditions are not degraded by invasive species, herbivory, fire, or other disturbances.

Fiddleheads in a spring forest.

Photo by Maria K. Janowiak, U.S. Forest Service and Northern Institute of Applied Climate Science

Additional considerations for individual forest types under this approach include:

- **Lowland Conifer, Lowland Hardwood:** Lowland systems may be at risk from drier conditions and encroachment of upland species. At the same time, options for maintaining these types under climate change are limited by the need for specific hydrologic conditions. Identifying and establishing refugia will likely provide the best opportunity for maintaining lowland systems on the landscape because few adaptation approaches will be effective in this forest type. Maintaining these systems as refugia may require active management of water levels and species composition.

Prioritize and protect sensitive or at-risk species or communities

Northern and boreal species are widespread in northern Wisconsin, but are likely to lose habitat because they are already at the southern extent of their range (Swanston et al. 2011). By prioritizing maintenance of sensitive and at-risk species or communities, managers can sustain these species on-site for as long as possible or until new long-term sites can be located and populated. An example of an adaptation tactic under this approach is to identify and protect high-quality stands of hemlock or other desired forest types to serve as refugia identified for long-term maintenance of the type. This approach could also be used to identify and create refugia for threatened and endangered plant or animal species.

Additional considerations for individual forest types under this approach include:

- **Aspen, Balsam Fir, Hemlock, Paper Birch, Spruce:** Many species in these forest types are projected to decline as a result of climate change; identifying refugia may be particularly important for maintaining these types on the landscape. Cooler and wetter sites such as riparian areas, north-facing slopes, lake edges, and wetlands may have suitable site conditions and may be less likely to be impacted by human-caused or natural disturbances.

- **Northern Hardwood:** Higher species diversity in this forest type increases its overall adaptive capacity, but several individual species are expected to decline. Actions to establish refugia on cooler and moister microhabitats may help maintain desired plant communities that are at risk from climate change.

Establish artificial reserves for at-risk and displaced species

Species already exist outside their natural habitats in nurseries, arboretums, greenhouses, and botanical gardens across the world. These artificial reserves may be used to support individuals or genetic lineages that are no longer able to survive in their former location (Fiedler and Laven 1996, Millar 1991). These highly controlled environments could act as interim refugia for rare and endangered plant species that have specialized environmental requirements and low genetic diversity (Heinz Center 2008, Spittlehouse and Stewart 2003). This idea of interim refugia to maintain species until they can be moved to new locations would likely require substantial resources to pursue (Coates and Dixon 2007). An example of an adaptation tactic under this approach is to use an existing artificial reserve to cultivate southern species whose suitable habitat has moved northward, but who face considerable lag time before new habitat may become available.

Strategy 5: Maintain and enhance species and structural diversity

Species and structural diversity may buffer a system against the susceptibility of individual components to changes in climate. Forest managers already work to increase structural and species diversity; as an adaptation strategy, this general goal receives added effort and focus. A system may still experience stress as individual components fare poorly, but the overall system can be made more resilient.

Approaches:

Promote diverse age classes

Species are vulnerable to stressors at different stages in the species life cycle. Maintaining multiple age classes of a species will help increase structural diversity within stands or across a landscape, as well as buffer vulnerability to stressors of any single age class. Monocultures and even-aged stands are often more susceptible to insect pests and diseases, many of which are likely to increase in range and severity as a result of climate change; maintaining a mosaic of even-aged stands of varying ages across the landscape will increase diversity in these forest types.

Maintain and restore diversity of native tree species

Diverse forests may be less vulnerable to climate change impacts because they distribute risk among multiple species, reducing the likelihood that the entire system will decline even if one or more species suffers adverse effects. This relationship may be especially important in forest types with low diversity; even small increases in diversity may increase resilience without greatly altering species composition or successional stage. Climate change may exacerbate adult mortality or induce regeneration failure of northern species (Swanston et al. 2011). Actions to promote and enhance regeneration of native species through understory management and planting efforts may help to maintain diverse and vigorous native communities.

Retain biological legacies

Biological legacies of desired species can facilitate persistence, colonization, adaptation, and migration responses to climate change (Gunn et al. 2009). Silvicultural treatments designed to retain biological legacies can be conducted to create diversity in structure, species composition, and unique characteristics while maintaining the appropriate density of desired species. An example of a tactic that is already in practice in northern Wisconsin is to retain individual trees of a variety of species to maintain their presence on the landscape. This tactic could also be used to provide both a potential seed source for species and genotypes that are expected to be better adapted to future conditions, as well as future nurse logs for regeneration of some species (Gunn et al. 2009).

Restore fire to fire-adapted ecosystems

Long-term fire suppression leads to shifts in forest structure and composition, which may disproportionately favor a smaller number of species and reduce biodiversity (Tirmenstein 1991). Restoring fire regimes that attempt to mimic natural disturbance in fire-adapted systems can enhance regeneration and encourage stronger competition by fire-dependent and fire-tolerant species (Abrams 1992). Repeated low-intensity fire in some forest types, such as red pine and oak, can emulate natural processes to foster more complex stand structures while reducing risk of severe fire. An example of an adaptation tactic under this approach is to use prescribed fire to reduce ladder fuels and lower risk of large and severe wildfire in areas that are expected to have increased fire risk as a result of climate change.

Establish reserves to protect ecosystem diversity

Some areas with exemplary combinations of soil, hydrologic, and climatic variation support a correspondingly high degree of species diversity. Ecosystem diversity in these areas may be protected by establishing reserves, traditionally defined as natural areas with little to no harvest activity that do not exclude fire management or other natural disturbance processes (Halpin 1997). The use and definition of reserves should be considered carefully within the context of changing climate and forest response as some systems may greatly benefit from minimal intrusion, whereas others may actually require more active management if ecosystem integrity begins to deteriorate. It may be valuable to retain explicit flexibility in management practices, so long as management directly supports the justifications and goals for establishing the reserve.

Additional considerations for individual forest types under this approach include:

- **Aspen, Jack Pine, Paper Birch:** This approach may not be well-suited to these forest types because they require frequent and active management to be maintained in an early-successional stage.

Strategy 6: Increase ecosystem redundancy across the landscape

Some losses are inevitable, whether due to catastrophic events or unforeseen interactions of management, climate change, and forest response. Increasing ecosystem redundancy attempts to lower the overall risk of losing a forest type by increasing the extent of the forest type, the number of occurrences of the forest type across the landscape, and the diversity of regeneration stages.

Approaches:

Manage habitats over a range of sites and conditions

The suitable site conditions for a forest type or species may shift on the landscape as climate changes. Spreading forest types over a range of sites and conditions, both existing and new, will increase combinations of location, site conditions, and species aggregations. Opportunities for successful regeneration and the likelihood of persistence of a

A managed jack pine stand.

Photo by Maria K. Janowiak, U.S. Forest Service and Northern Institute of Applied Climate Science

species or community may thus be increased (Joyce et al. 2009, Millar et al. 2007, TNC 2009).

Additional considerations for individual forest types under this approach include:

- **Aspen, Jack Pine, Northern Hardwood:** These forest types are currently widespread across a range of sites and conditions, but expected to decline. The current extent may provide many options for retaining redundancy across the landscape.

Expand the boundaries of reserves to increase diversity

Certain areas contain exemplary combinations of soil, hydrologic, and climatic variation, with a correspondingly high degree of species diversity. An earlier approach (Establish reserves to protect ecosystem diversity) describes choosing areas such as these to establish reserves, which are traditionally defined as natural areas with little to no harvest activity that do not exclude fire management or other natural disturbance processes (Halpin 1997). Expanding existing reserve boundaries may buffer and replicate the diversity within the core of the reserve, but more importantly, may also increase the overall variation in species within the expanded reserve.

Strategy 7: Promote landscape connectivity

Species migration is a critical factor in the ability of forests to maintain ecosystem function in a changing climate; however, fragmentation of landscapes and loss of habitat may restrict species movements and gene flow (Davis and Shaw 2001). Managing the landscape for connectivity will allow for easier movement, reduce lags in migration, and enhance genetic diversity. This approach benefits both wildlife and forests (e.g., squirrels provide the movement of acorns across the landscape). Connectivity also increases movement of invasives and pests, however, thereby increasing the need for

vigilance. Numerous factors contribute to today's migration patterns. The current rate of climate change, coupled with contemporary land use and demographics, creates challenges to migration that are unique to this period in time. Many species are not expected to be able to migrate at a rate sufficient to keep up with climate change and associated range shifts (Davis and Shaw 2001, Iverson et al. 2004), so when managers pursue this strategy, it may be beneficial to combine the approaches under this strategy with ones to create refugia (Strategy 4).

Approaches:

Use landscape-scale planning and partnerships to reduce fragmentation and enhance connectivity

Enabling species migration across the landscape and buffering against disturbance will require a concerted effort to create partnerships, agreements, and other mechanisms for land protection and management across property boundaries. Coordinating forest conservation easements and certification programs, and other efforts to increase the size and connectivity of forests will foster a landscape-level response to counter the widespread effects of climate change (Millar et al. 2007, Spittlehouse and Stewart 2003). Establishing management agreements across boundaries of reserves and managed forests will also allow for protection of species moving across the landscape (Heller and Zavaleta 2009).

Establish and expand reserves and reserve networks to link habitats and protect key communities

Areas containing exemplary combinations of soil, hydrologic, and climatic variation may have a correspondingly high degree of species diversity. An earlier approach (Establish reserves to protect ecosystem diversity) describes choosing these areas to establish reserves, traditionally defined as natural areas with little to no harvest activity that do not exclude fire management or other natural disturbance processes (Halpin 1997). Placing reserves adjacent to each other to form a network of a few large reserves, many small reserves along a latitudinal gradient, or a combination of large and small reserves close to each other will help maintain connectivity across a varied and dynamic landscape (Halpin 1997, Heller and Zavaleta 2009, Spittlehouse and Stewart 2003). Designating buffer zones of low-intensity management around core reserve areas and between different land uses may allow greater flexibility in silvicultural treatments of adjacent lands, promote species movement, and help protect core areas from disturbance.

Maintain and create habitat corridors through reforestation or restoration

The presence of both small and large corridors on the landscape may help slow-moving species to migrate without additional assistance. Establishing or restoring forest cover along natural features, such as rivers, or property lines may improve species' ability to naturally adapt and migrate (Heller and Zavaleta 2009). Corridors oriented in any direction may be useful to facilitate genetic mixing, but corridors arranged north-south may be more useful if the goal is to allow for species movements along the climatic gradient.

Additional considerations for individual forest types under this approach include:

- **Lowland Conifer, Lowland Hardwood:** Climate change is likely to alter the hydrologic conditions required by these forest types. Reforestation or restoration of riparian areas may help retain these species on the landscape longer while providing a forested corridor.

- **Northern Hardwood:** This forest type could serve as a matrix to connect many forest types across the landscape because it is currently abundant and relatively contiguous in northern Wisconsin. These forests can be used as migration routes for southern species to move northward.

A northern flicker on the branch of a dead tree.

Photo by Maria K. Janowiak, U.S. Forest Service and Northern Institute of Applied Climate Science

Strategy 8: Enhance genetic diversity

Climate has changed in the past, but the current rate of climate change is much more rapid than in the past (Davis and Shaw 2001, Woodall et al. 2009), and human dependence on ecosystem services may demand an active role in species migration and genetic adaptation (McLachlan et al. 2007). Greater genetic diversity within and among species may help species adjust to new conditions or sites, distributing the risk across a more diverse population. Actions to enhance genetic diversity could be timed to occur after large-scale disturbances to take advantage of regeneration and establishment phases. Approaches under this strategy are best implemented with great caution, incorporating due consideration of the uncertainties inherent in climate change, the sparse record of previous examples, and continued uncertainties of forest response. Initial attempts at implementation may be considered as early field trials rather than templates.

Approaches:

Use seeds, germplasm, and other genetic material from across a greater geographic range

Many key environmental factors that influence plant growth and reproduction are changing, including temperature, length of growing season, and the amount and seasonal pattern of precipitation (Post 2003). Out-planted nursery seedlings typically have greater survival when they originate from local seed sources, but these seed sources may no longer produce the best-adapted seedlings if the governing environmental factors change. Utilizing seed zones that change over time and are based on regional analyses of climate change data may provide better seed sources than static seed zones (Millar et al. 2007, Spittlehouse and Stewart 2003). This approach may entail importing seedlings from slightly farther away that are better adapted to current or future climatological conditions. Genetic sources may be limited for the many species that are currently at the southern edge of their range in northern Wisconsin.

For other species, new genetic material may be available from sources to the south or west. It is important to note that although many environmental factors may match seedlings to geographic area, cold tolerance or other limitations may remain (Millar et al. 2007).

Favor existing genotypes that are better adapted to future conditions

Genotypes may be present within forests that are better adapted to future conditions because of pest resistance, broad tolerances, or other characteristics (Millar et al. 2007, Spittlehouse and Stewart 2003). However, the use of this approach may be currently limited by the uncertainty surrounding precise future conditions and which genotypes are best suited to these conditions. The use of this approach may also be limited by available source material. An example of an adaptation approach under this tactic is to collect seed from vigorous and healthy trees in locations that currently have conditions that are expected to be more prevalent in the future. Another example is to retain some vigorous survivors of a die-back event, such as drought-induced mortality, rather than salvage all trees in an affected area.

Increase diversity of nursery stock to provide those species or genotypes likely to succeed

Maintaining ecosystem function and diversity is largely dependent on successful seedling establishment, which may require historically unprecedented planting efforts in some locations. Changing climatic conditions will need to be paralleled by appropriate infrastructure and resources for regeneration, including the availability of genetically diverse material coming from seed orchards and nurseries (Millar et al. 2007). Because infrastructure and resources to support the development of new sources of stock may be limited, it may become more important to invest in nurseries that provide an array of species and genotypes that can both meet short-term demand for traditional species and enable long-term adaptation.

Strategy 9: Facilitate community adjustments through species transitions

Species composition in many ecosystems is expected to change as species adapt to a new climate and form new communities. This strategy seeks to maintain overall ecosystem function and health by gradually enabling and assisting adaptive transitions within forest communities. The result may be species assemblages slightly different from those present in the current forest type, or an altogether different forest type in future decades. Importantly, this strategy aims to maintain key ecosystem functions, not an unchanging community or species mix. Further, this strategy is unlikely to be applicable across the entire range of any forest type within the next several decades. Many of the approaches in this strategy attempt to mimic natural processes, but may currently be considered unconventional management responses. In particular, some approaches incorporate assisted migration, which remains a challenging and contentious issue (Janowiak et al. 2011, McLachlan et al. 2007). Caution is warranted in use of this strategy, and initial attempts at implementation may serve more as early field trials than as templates. Outcomes from early efforts to promote community transition can be evaluated to provide information on future opportunities for transitions, as well as effective methods and timing.

Approaches:

Anticipate and respond to species decline
Species on the southern and warmer edges of their geographic range are especially vulnerable to habitat loss, and some systems are expected to decline rapidly as conditions change. In northern Wisconsin, nearly all forest types contain dominant or common associate species that are expected to have substantial declines in suitable habitat under a changing climate (Swanston et al. 2011). Anticipating forest and species declines due to changes in climate, disturbance regimes, or other factors may help in developing early and appropriate management responses to maintain forest cover

and ecosystem function. The species assemblage may be dramatically altered through active or passive means. An example of an adaptation tactic under this approach is to identify tree species that are very likely to decline or are already declining, and promote other species to fill a similar niche. For example, red maple and black cherry could be encouraged on a site that is expected to become drier to compensate for expected decreases in sugar maple dominance.

Additional considerations for individual forest types under this approach include:

- **Lowland Hardwood:** Lowland species are likely to be impacted by altered hydrology and precipitation patterns as a result of climate change (Swanston et al. 2011). Susceptibility to the emerald ash borer, however, may result in complete loss of forest cover in black ash-dominated sites long before climate change impacts are evident. Early promotion of other lowland hardwood species may buffer the decline of ash-dominated stands and help retain ecosystem function in the short term.

Favor or restore native species that are expected to be better adapted to future conditions
In many cases, native species may be well-adapted to the future range of climatic and site conditions. Using management to favor the native species in a forest type that are expected to fare better under future climate change can facilitate a shift in the species assemblage without drastically altering the forest composition. Although many forest types contain one or several dominant species that are expected to experience declines in suitable habitat (Swanston et al. 2011), other tree species may be able to be favored for their expected ability to do well under future conditions, such as red oak in the aspen, jack pine, and northern hardwood forest types. Where forests are dominated by a single species, this approach will likely lead to conversion to a different forest type, albeit with a native species.

Photo by Maria K. Janowiak, U.S. Forest Service and Northern Institute of Applied Climate Science

A stream winding through a northern hardwood forest.

Manage for species and genotypes with wide moisture and temperature tolerances

Inherent scientific uncertainty surrounds climate projections at finer spatial scales (Schiermeier 2010), making it necessary to base decisions upon a wide range of predictions of future climate. Managing for a variety of species and genotypes with a wide range of moisture and temperature tolerances may better distribute risk than attempting to select species with a narrow range of tolerances that are best adapted to a specific set of future climate conditions (TNC 2009).

Emphasize drought- and heat-tolerant species and populations

Some areas of northern Wisconsin have already experienced decreased summer precipitation over the last half century (Kucharik et al. 2010, WICCI

2011b). There is uncertainty about whether this trend will continue, but expected increases in large precipitation events may concentrate rainfall to fewer total events (Swanston et al. 2011, WICCI 2011b). In anticipation of warmer temperatures and potential for drought late in the growing season, it may be beneficial to emphasize drought- and heat-tolerant species in areas that may be most vulnerable to drought. An example of an adaptation tactic under this approach is to favor or establish oak species on narrow ridge tops, south-facing slopes with shallow soils, or other sites that are expected to become warmer and drier. Another example is to seed or plant drought-resistant genotypes of commercial species where there is an expectation of increased drought stress (Joyce et al. 2009).

Additional considerations for individual forest types under this approach include:

- **Aspen, Northern Hardwood:** Actions to promote oak, pine, and other more drought- and heat-tolerant species on drier sites where these species are already present as a minor component may help to reduce the vulnerability of these forest types.

- **Jack Pine:** This type is the most drought-tolerant forest type present in northern Wisconsin. While jack pine is projected to experience declines under future climate conditions, the species may fare better than the models predict (Swanston et al. 2011). Jack pine forest may be able to persist on many sites, including sites that currently do not host jack pine but may become more suitable in the future.

Guide species composition at early stages of stand development

Long-term ecosystem function may be jeopardized if existing and newly migrated species fail to regenerate and establish. Active management of understory regeneration may help forests make the transition to new and better-adapted compositions more quickly by reducing competition from undesirable, poorly-adapted, or invasive species. Natural disturbances often initiate increased seedling development and genetic mixing, and can be used to facilitate adaptation (Joyce et al. 2009). When forests are dominated by one or a small number of species, this approach may lead to conversion to a different forest type.

Additional considerations for individual forest types under this approach include:

- **Jack Pine, Paper Birch, Red Pine:** Under drier conditions and increased stress, promoting regeneration and discouraging competitors may require more intensive site preparation, including prescribed fire, soil disturbance, and herbicide use.

Protect future-adapted regeneration from herbivory

Herbivory from insects, rodents, ungulates, and other species may increase as a result of climate change (WICCI 2011b), potentially interacting with other agents to increase overall stress and reduce regeneration. Protecting desired regeneration of existing or newly migrated species can strongly shape the ways in which communities adapt (TNC 2009). In northern Wisconsin, herbivory from white-tailed deer is already a major determinant of species composition through direct effects on regeneration success (Waller and Alverson 1997, Waller 2007, WDNR 2010); therefore, efforts to counter herbivory may focus on deer in many locations. Examples of adaptation tactics under this approach include the use of repellent sprays, bud caps, or fencing to prevent browse on species that are expected to be well-adapted to future conditions.

Establish or encourage new mixes of native species

Novel combinations of climatic and site conditions may support mixtures of native species that do not currently occur together. While some species may not typically occur in the same forest type, they may have been together previously (e.g., butternut and American elm in the oak forest type; Rhemtulla et al. 2009). Novel mixing of native species may result in conversion to a newly defined or redefined forest type. This approach is best implemented with great caution, incorporating due consideration of the uncertainties inherent in climate change, the sparse record of previous examples, and continued uncertainties of forest response. Initial attempts at implementation may be considered as early field trials rather than templates.

Identify and move species to sites that are likely to provide future habitat

Climate may be changing more rapidly than some tree species can migrate, and the northward movement of species may be restricted by land use or other impediments between areas of suitable habitat (Davis and Shaw 2001, Iverson et al. 2004). Maintaining ecosystem function or making the transition to a better-adapted system may involve the active introduction of nonnative species or genotypes (McLachlan et al. 2007). Given the uncertainty about specific climate conditions in the future, the likelihood of success may be increased by relocating species with a broad range of tolerances (e.g., temperature, moisture) across a wide range of provenances. This approach is best implemented with great caution, incorporating due consideration of the uncertainties inherent in climate change, the sparse record of previous examples, and continued uncertainties about forest response. Initial attempts at implementation may serve more as early field trials than as templates.

Strategy 10: Plan for and respond to disturbance

Ecosystems may face dramatic impacts as a result of climate change-related alterations in disturbances, including fire, drought, invasive species, and severe weather events (Dale et al. 2001). Disturbances are primary drivers of some ecosystem dynamics (e.g., stand-replacing fire in jack pine and windthrow in northern hardwood forests; Johnson 1995), but changes in the frequency, intensity, and duration of disturbance events may create significant management challenges. Although it is not possible to predict and prepare for a single disturbance event, it is possible to increase overall preparedness for large and severe disturbances. Actions to respond to disturbance can take place in advance of, as part of, or following the event (Dale et al. 2001). Many of the best opportunities for addressing disturbance-related impacts may occur immediately after the disturbance event; having a suite of planned options in place may facilitate an earlier and more flexible response. This strategy asks forest managers to imagine the worst-case scenarios, consider new opportunities, and have plans in place to quickly and appropriately respond.

Approaches:

Prepare for more frequent and more severe disturbances

Disturbances are likely to occur outside historical patterns, and impacts on forest ecosystems may occur more frequently and be more severe. Documenting clear plans for how to respond to more frequent or severe disturbances in advance will allow for a faster, more thoughtful, better-coordinated response. An example of an adaptation tactic under this approach is to identify locations where a given forest type would be unlikely to successfully reestablish in the case of a severe disturbance event, and then develop response options for establishing better-adapted communities in these places should the event occur.

Additional considerations for individual forest types under this approach include:

- **Aspen, Paper Birch:** Stand-replacing disturbances are often beneficial to these forest types. However, these early-successional forest types may experience accelerated succession where small- and medium-scale disturbances break up areas of forest canopy to allow shade-tolerant species to establish.

- **Balsam Fir, Spruce:** The dominant species in these forest types have flammable needles and shallow roots, which make them highly susceptible to damage or mortality from fire and wind.

- **Jack Pine:** Jack pine forests require extensive site disturbance to regenerate and are generally favored by stand-replacing disturbances and recurring fires. However, drought or fire can kill young seedlings and untimely disturbances may interfere with regeneration.

- **Lowland Conifer:** Increased temperatures and reduced seasonal precipitation may dry wet soils and peatlands, leading to increased risk of wildfire.

Prepare to realign management of significantly altered ecosystems to meet expected future environmental conditions

Some ecosystems may experience significant disruption and decline, such that desired conditions or management objectives may no longer be feasible. Management of these systems may be realigned to create necessary changes in species composition and structure to better adapt forests to current and anticipated environments, rather than historical pre-disturbance conditions (Millar et al. 2007, Spittlehouse and Stewart 2003). Developing clear plans that establish processes for realigning significantly altered ecosystems before engaging in active management will allow for more thoughtful discussion and better coordination with other adaptation responses. An example of realignment that is currently occurring in northern Wisconsin is the replacement of failed spruce forests with jack pine, tamarack, and other species.

Promptly revegetate sites after disturbance

Potential increases in the frequency, intensity, and extent of large and severe disturbances may disrupt regeneration and result in loss of forest cover, productivity, or function in the long term. Prompt revegetation of sites following disturbance may help reduce soil loss and erosion, protect water quality, discourage invasive species, and improve aesthetic quality in the newly exposed areas.

Allow for areas of natural regeneration after disturbance

Although many areas may undergo site preparation or be replanted after severe disturbance, some areas can be set aside to allow for natural regeneration as a means to identify the well-adapted species and populations (Joyce et al. 2009). The use and monitoring of test or "control" areas of natural revegetation following disturbance may help inform managers regarding species that can successfully regenerate at a local level without intervention.

Maintain seed or nursery stock of desired species for use following severe disturbance

Disturbed areas may need to be reseeded or replanted with high quality, genetically appropriate, and diverse stock. Maintaining stock that represents a variety of environmental conditions across a broad geographic range may help to supply future-adapted species and genotypes when needed. Advanced planning can help to make materials more available and improve the ability to promptly respond to disturbances. This approach may be limited by the availability of infrastructure and resources. There is also uncertainty regarding which genotypes may be better suited to future conditions, and the development of new sources of stock may take decades.

Remove or prevent establishment of invasives and other competitors following disturbance

Disturbed sites are more susceptible to colonization by invasive species, which are expected to increase under climate change (Dukes et al. 2009, Hellman et al. 2008) and may outcompete regeneration of desired species (Joyce et al. 2009). Emphasizing early detection of and rapid response to new infestations may reduce competition and aid efforts to encourage successful regeneration of desired species. Nonnative species that have desirable characteristics and are not invasive may be a lower priority for removal, perhaps becoming desirable ("neo-native") species in the future (Millar and Brubaker 2006, Millar et al. 2007).

CHAPTER 3: ADAPTATION WORKBOOK

Maria Janowiak, Patricia Butler, Chris Swanston, Linda Parker,
Matt St. Pierre, and Leslie Brandt

Climate change imposes many challenges on the long-term management of ecosystems and is becoming an increasingly important consideration in land management planning and decisionmaking at a variety of spatial scales. The process outlined in this chapter helps managers incorporate climate change considerations into management planning and activities, while complementing existing processes and procedures for making decisions (Box 5). Moreover, it uses a workbook method to provide instructions for land managers to translate adaptation strategies and approaches (Chapter 2) into management tactics that can help forest ecosystems adapt to climate change.

About the Adaptation Workbook

As more information becomes available about the expected effects of climate change, natural resource managers are placing greater emphasis on responding to climate change. However, the amount of information that has been specifically developed to help land managers make on-the-ground management decisions is currently limited. Given the uncertainty around future conditions at specific locations or points in time, the most substantial challenge of managing forests in the face of climate change may be to begin using the tools and information currently at hand (Janowiak

---

Box 5: Using the Adaptation Workbook

The Adaptation Workbook can:

- Help managers view climate change as an emerging management consideration that can be incorporated into many aspects of existing management planning and decisionmaking.

- Integrate a wide variety of adaptation strategies and approaches into management decisions based upon existing management goals and objectives.

- Provide a platform for discussion of climate change-related topics and issues with co-workers, team members, and other collaborators.

- Document considerations and decisions regarding climate change and management.

The Adaptation Workbook does not:

- Make recommendations or set guidelines for management decisions or actions.

- Establish a plan for implementation of the selected tactics and monitoring efforts. Rather, that step is reserved for managers to pursue after completion of the workbook.

et al. 2011, Lawler et al. 2010). For this reason, an iterative adaptive management process (e.g., Stankey et al. 2005) that incorporates monitoring and the re-evaluation of management goals is well-suited to climate change adaptation activities (Lawler et al. 2010).

To meet this need, we have developed the Adaptation Workbook to help forest managers more effectively bring climate change considerations to the spatial and temporal scales where management decisions are made. The workbook is an analytical process built upon a conceptual model for adaptation (Fig. 7) that was derived from adaptive management principles. It has been designed to draw upon regionally specific information. Resources developed as part of the Climate Change Response Framework project in northern Wisconsin include the *Ecosystem Vulnerability Assessment and Synthesis* (Swanston et al. 2011) and the previous Adaptation Strategies and Approaches chapter (Chapter 2) of this document.

The workbook features a five-step process that can be used to incorporate climate change into resource management at a variety of spatial scales (e.g., stand, large-ownership) and at many levels of decisionmaking (e.g., planning, problem solving, implementation). By defining current management goals and objectives in the first step, the process is designed to integrate climate change adaptation into existing management efforts and consider what actions may be useful for responding to climate change. It is not intended to provide specific guidance or replace other forms of management planning; rather, it relies on the experience and expertise of natural resource professionals and is meant to complement existing management planning and decisionmaking systems[1]. The workbook is designed for flexibility and can be used for a wide

---

[1]The Adaptation Workbook is designed to supplement and support existing decisionmaking processes in the Forest Service and other agencies or institutions, but does not in any way replace, supersede, or circumvent those processes.

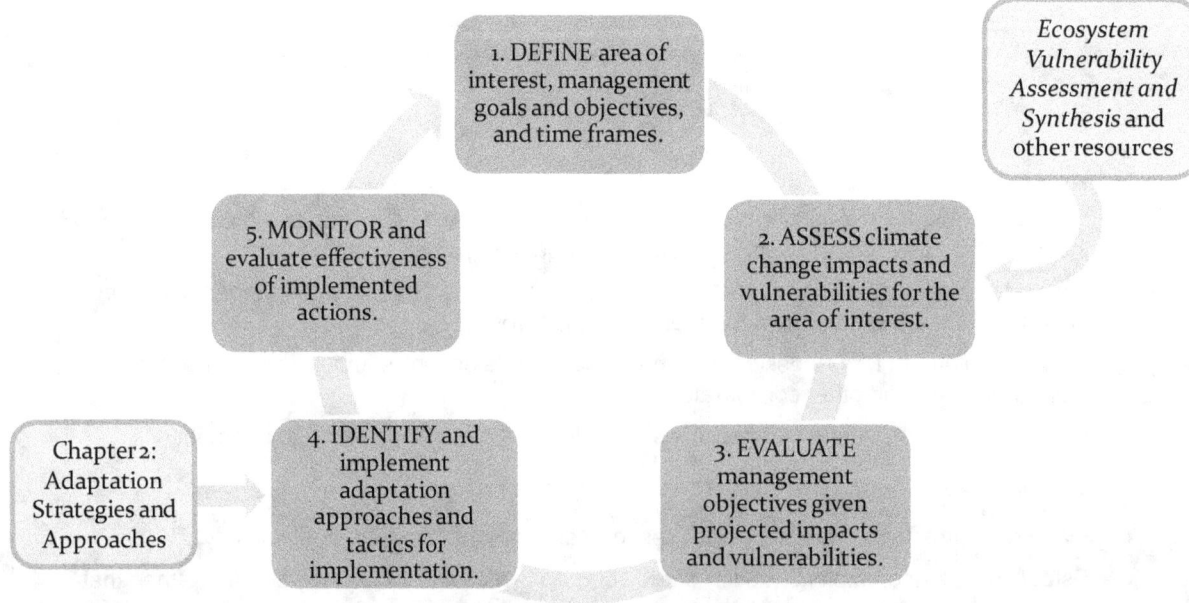

Figure 7.—The Adaptation Workbook presents a five-step process (dark green rectangles) that can be used to incorporate climate change as a management consideration and help ecosystems adapt to the anticipated effects of climate change. Additional resources (light green rectangles) provide information and tools that support the process.

range of applications; however, it is probably most useful for management activities at small to medium scales of management.

Adaptation Workbook Instructions

Welcome to the Adaptation Workbook! This process is designed to help you consider climate change adaptation in your resource management activities. The process consists of five sequential steps (Fig. 7); each step is described in detail with a corresponding table for recording information. A certain amount of flexibility is built into the Workbook; you can adjust it to include more or less information if desired.

When using this workbook, land managers in northern Wisconsin can draw upon information from specific resources such as the Adaptation Strategies and Approaches chapter (Chapter 2). Because the understanding of the impacts of climate change is continually changing and growing, we recommend that you consult recent information and resources that are relevant to your area of interest. We have noted several resources where appropriate. Additionally, we encourage you to consult with scientists and others with expertise on climate change impacts in your area of interest.

To get an overview of the process before you begin the workbook, we recommend that you read the entire Adaptation Workbook as well as the Illustrations in Chapter 4. Collect any needed information, including relevant maps and management plans, and then start the workbook.

Tables have been provided in the workbook to show how information is arranged, and electronic spreadsheets can be used for capturing your notes as you use the workbook.

The workbook also uses forest types as a way to group information (see Chapter 2 for regional forest type descriptions). If you are working in multiple forest types, you may find it easier to complete the process for each forest type individually during some steps of the workbook.

Integration with Other Chapters

The workbook is designed to be used with the Adaptation Strategies and Approaches outlined in the previous chapter or other adaptation strategies and approaches deemed relevant.

- In this workbook, you will use the Adaptation Strategies and Approaches from the previous chapter to select the strategies and approaches that will help you meet your management goals and objectives, given projected changes to the climate and ecosystems. Then, you will develop more specific actions (tactics) to implement your selected approach.
- The Illustrations chapter (Chapter 4) provides two examples of how the Adaptation Workbook was used with the Adaptation Strategies and Approaches to identify management tactics for adapting forest ecosystems to climate change. You may also find it helpful to refer to the Illustrations as you complete the workbook.

Step #1: DEFINE area of interest, management goals and objectives, and time frames.

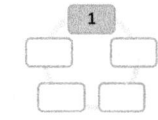

About this Step

In this step, you will write and record basic information about the area of interest, which can be either a specific location (geographic area) or an issue (topic area). It is likely that you already have this information available to you; this step is intended to record this information and help you to keep it in mind during subsequent steps (Worksheet #1, below; full-size worksheets for the user to cut out or copy are located in the back of this document following page 121).

Depending on what you've selected as your area of interest, you might complete this step in different ways. For example,

- You may want to define the geographic location first and then describe the management goals and objectives for that particular location. This may make more sense if you are working in a relatively small geographic area that is clearly defined (e.g., a particular stand or project area).

- Alternatively, you may prefer to define the management goals and objectives first and then the location or locations where those apply. This may make more sense if you are working at a larger spatial extent, such as a large ownership, or focusing on a specific management goal (e.g., habitat for a single species).

Worksheet #1. Full-size worksheets are located in the back of this document.

Area of Interest	Location	Forest Type(s)	Management Goals	Management Objectives	Time Frames

A D A P T A T I O N   W O R K B O O K

Description of Worksheet #1 Items

Area of Interest – Describe the area of interest, which can be either a geographic area (such as a management unit) or a topic area (such as a particular feature that is being managed for).

Location – Describe the geographic location or locations (e.g., stands, management units, or other information) for the area of interest.

Current Forest Type(s) – List the existing forest types (refer to list in Chapter 2: Adaptation Strategies and Approaches) that are relevant to your location.

Note – If desired, describe current conditions or desired future conditions in more detail.

Note – If the management goals or planned activities include the conversion of current forest types to another forest type, also list the desired future forest type (and conditions, if desired).

Management Goals – List the management goals for the area of interest (Box 6). Management goals may include the desired future forest types or conditions, habitat characteristics, the production of products, or other ecological features or services.

Management Objectives – List any management objectives for the area of interest (see Box 6). These will explain how management goals will be achieved, and there may be multiple objectives for a single management goal.

Time Frames – List approximate time frames for implementing management actions and for achieving management goals and objectives. Management horizons often span decades to incorporate all aspects of assessment, implementation, monitoring, and evaluation. For example, a management action such as a harvest may be planned to occur within 10 years (short-term) as a means to achieve an objective. A corresponding management goal related

---

Box 6: Goals and Objectives

Management Goals

Management goals are broad, general statements, usually not quantifiable, that express a desired state or process to be achieved (Society of American Foresters 2011). They are often not attainable in the short term, and they provide the context for more specific objectives. Examples of management goals are:

- Maintain and improve forest health and vigor.
- Maintain wildlife habitat for a variety of species.

Management Objectives

Management objectives are concise statements of measurable planned results that correspond to preestablished goals in achieving a desired outcome (Society of American Foresters 2011). These objectives commonly include information on resources to be used and form the basis for further planning to define the precise steps to be taken to achieve the identified goals. Examples of management objectives include:

- Regenerate a portion of the oldest aspen forest type through clearcut harvest in the next year in order to maintain and improve forest health and vigor in young aspen stands.
- Implement silvicultural treatments within 5 years in order to increase the oak component of selected stands and enhance wildlife habitat.

to species composition or forest structure may have a time frame of 30 years or more (long-term), with immediate and short-term actions nested within that time frame. As a default, use the following categories to identify your relevant time frame(s), but allow the objectives and ecosystems to dictate the most appropriate time frames. As you progress through the workbook, feel free to revisit and adjust them as necessary:

- Immediate: 2 years or less
- Short-term: 2-10 years
- Medium-term: 10-30 years
- Long-term: 30 or more years

National Forest managers discussing management options in an aspen stand.

Photo by Maria K. Janowiak, U.S. Forest Service and Northern Institute of Applied Climate Science

Step #2: ASSESS climate change impacts and vulnerabilities for the area of interest.

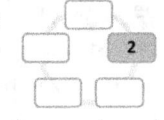

About this Step

Identifying the ecosystem components that are most vulnerable is a critical first step in considering climate change impacts and developing responses to help systems adapt (Glick et al. 2011, National Research Council 2010). Careful consideration of impacts across vulnerable and resilient ecosystems can help you prioritize management responses for the areas where the largest impacts are likely to occur.

In this step, you will draw upon the ever-growing body of information about the projected effects of climate change as well as your expertise and experience in order to more specifically assess ecosystem vulnerability to a range of projected climate change. Because there is a lot of variability among different locations, your understanding of specific site conditions in the area of interest will help tailor management responses in this and later steps (Worksheet #2, below).

We recommend that you review the *Ecosystem Vulnerability Assessment and Synthesis* (Swanston et al. 2011) in this step because it summarizes impacts to forests in northern Wisconsin. Additionally, other

Worksheet #2. Full-size worksheets are located in the back of this document.

Broad-scale Impacts and Vulnerabilities	Climate Change Impacts and Vulnerabilities for the Area of Interest	Vulnerability Determination
	How might broad-scale impacts and vulnerabilities be affected by conditions in <u>your area of interest?</u> • Landscape pattern • Site location, such as topographic position or proximity to water features • Soil characteristics • Management history or current management plans • Species or structural composition • Presence of or susceptibility to pests, disease, or nonnative species that may become more problematic under future climate conditions • Other….	

information can be included to supplement that assessment. The use of multiple information sources during this step will provide a greater amount of background on anticipated climate change effects.

Description of Worksheet #2 Items

Broad-scale Impacts and Vulnerabilities – List climate change impacts and vulnerabilities for the region that you are working in (e.g., northern Wisconsin) as well as the source of this information. These may be specific to the forest type(s) that you have listed. As a starting point, broad-scale impacts and vulnerabilities for northern Wisconsin were

described in the *Ecosystem Vulnerability Assessment and Synthesis* (Swanston et al. 2011) and have been summarized below as a starting point for completing this item (Table 3). Additionally, many resources on climate change impacts and vulnerabilities exist, such as reports and peer-reviewed papers on climate change (see Box 7 for a short list of resources).

Climate Change Impacts and Vulnerabilities for the Area of Interest – The broad-scale impacts and vulnerabilities that you have listed apply generally across wide areas, but may be more or less important for your area of interest because of specific conditions or features associated with the

Table 3.—Climate change-related impacts and vulnerabilities for northern Wisconsin, from Swanston et al. (2011). The forest types presented in this table are defined in Table 1 of this document.

Extent	Potential climate change impacts and vulnerabilities for northern Wisconsin Forests
All forests in northern Wisconsin	Warmer temperatures Longer growing seasons Altered precipitation regimes Drier soils during summer Increased threats from insects, diseases, and invasive plants Altered disturbance regimes may lead to changes in successional trajectories Many common tree species are projected to have reduced habitat suitability Decline of associated rare species Decline of associated wildlife species
Aspen	Increased medium- and large-scale disturbances Decline of quaking aspen abundance or productivity Low within-stand diversity may increase risk of substantial aspen declines Medium-scale disturbances may not adequately allow for reestablishment Lack of genetic diversity within clones may be a likely disadvantage
Balsam fir	Habitat suitability may be substantially decreased Forest is less resilient to disturbances Increased competition with shade-tolerant species, such as red maple
Hemlock	More summer storms and wind events may lead to shifts in prevailing natural disturbance regimes Acceleration of current decline Drier conditions and increased disturbances may exacerbate current regeneration limitations Static ecosystem is less resilient to disturbance
Jack pine	Increased risk of fire occurrence Decline in productivity, especially on very dry sites
Lowland conifer	Altered hydrology and precipitation patterns may lead to reduced duration of soil saturation or ponding Increased risk of fire occurrence in dried organic soils Habitat suitability may be substantially decreased Reduced soil moisture or saturation may cause declines in hydrophytic tree species Static ecosystem is less resilient to disturbance

Table 3 (continued).—Climate change-related impacts and vulnerabilities for northern Wisconsin, from Swanston et al. (2011). The forest types presented in this table are defined in Table 1 of this document.

Extent	Potential climate change impacts and vulnerabilities for northern Wisconsin Forests
Lowland hardwood	Altered hydrology and precipitation patterns may lead to reduced duration of soil saturation or ponding Black ash habitat suitability may be substantially decreased Low within-stand diversity may increase risk if black ash declines substantially Drier conditions may lead to increased competition from other tree and plant species Emerald ash borer may interact with other stressors to cause widespread mortality
Northern hardwood	More summer storms and wind events may alter prevailing natural disturbance regimes Increased root damage from altered freeze-thaw cycles Decline of sugar maple productivity, especially on drier sites Increased disturbances may accelerate current decline of eastern hemlock and yellow birch Drying of ephemeral ponds may increase stress on associated species
Oak	Decline in productivity, especially on very dry sites
Paper birch	Increased fire and wind disturbance Increased disturbances may accelerate current decline Wind or other medium-scale disturbances may not adequately allow for reestablishment
Red pine	Increased risk of fire occurrence Low within-stand diversity may increase risk of substantial declines Younger stands may be vulnerable to pests that are currently present in warmer locations, especially under drought conditions Increased competition from some deciduous species, such as red maple and red oak
Spruce	Habitat suitability may be substantially decreased for white spruce and several associated species Drier soils may affect shallow-rooted white spruce Interactions among pests, drought, and other stressors may exacerbate current declines
White pine	Decline on drier sites due to drought-intolerance Super-canopy structure may increase individual tree mortality Increased competition from some associated species, such as red oak

area or ecosystem. For example, a site may have greater vulnerability to anticipated increases in the frequency and intensity of storm events because of its age structure or species composition, or less vulnerability to late-season moisture deficits due to a combination of topographic position and high water table.

Drawing upon your experience and knowledge of your area of interest, describe how you might expect the broad impacts and vulnerabilities to be modified in your area (see Box 8 for a list of considerations). You may want to reword the statements in the previous column to better reflect their more specific interaction with your area. Additionally, information from the other resources used to assess the broad-scale impacts and vulnerabilities may be useful in providing more information focused on the area of interest.

Vulnerability Determination – Vulnerability is the susceptibility of a system to the adverse effects of climate change. It is a function of its sensitivity to climatic changes, its exposure to those changes, and its ability to cope with climate change impacts with minimal disruption (Glick et al. 2011, Levina and Tirpak 2006; Fig. 8). For example, an ecosystem subject to few potential impacts and having a high adaptive capacity would be determined to have low vulnerability.

Box 7: Climate Change Effects

Impacts and Vulnerabilities

It is important to consider both climate change impacts and vulnerabilities in this step:

- *Impacts* integrate the degree of change (exposure) that a species or system is likely to experience and the likely response to change (sensitivity; Glick et al. 2011).
- *Vulnerability* is the susceptibility of a system to the adverse effects of climate change. Vulnerability is a function of its sensitivity to climatic changes, its exposure to those changes, and its capacity to adapt to those changes with minimal disruption (Glick et al. 2011; Levina and Tirpak 2006).

Sources of Climate Change Information

Many resources are available that provide information on climate change impacts and vulnerabilities in northern Wisconsin and the region:

- The *Ecosystem Vulnerability Assessment and Synthesis* (Swanston et al. 2011) summarizes information on climate change impacts, vulnerabilities, and expected changes for forested ecosystems in northern Wisconsin.
- The Wisconsin Initiative on Climate Change Impacts website (WICCI 2011a) contains information on projected changes in temperature, precipitation, and extreme weather for Wisconsin. Information on climate change impacts statewide were also compiled into an assessment (WICCI 2011b).
- *Scanning the Conservation Horizon: a Guide to Climate Change Vulnerability Assessment* (Glick et al. 2011) is a good starting resource for creating a new vulnerability assessment.
- The report, *Global Climate Change Impacts in the United States* (U.S. Global Change Research Program 2009), describes climate change impacts for all regions of the United States.
- The series, *Confronting Climate Change in the U.S. Midwest* (Union of Concerned Scientists 2009), describes the potential consequences of climate change for individual states in the region.

Box 8: Climate Change and Your Area of Interest

Most of the available information on climate change impacts has likely been developed for spatial scales that are larger than your area of interest. It is important to consider not only these broad-scale impacts, but also how they may be expressed in your particular area of interest. Factors that may alter broad-scale impacts include:

- Landscape pattern
- Site location, such as topographic position or proximity to water features
- Soil characteristics
- Management history or current management plans
- Species or structural composition
- Presence of or susceptibility to pests, disease, or nonnative species that may become more problematic under future climate conditions.

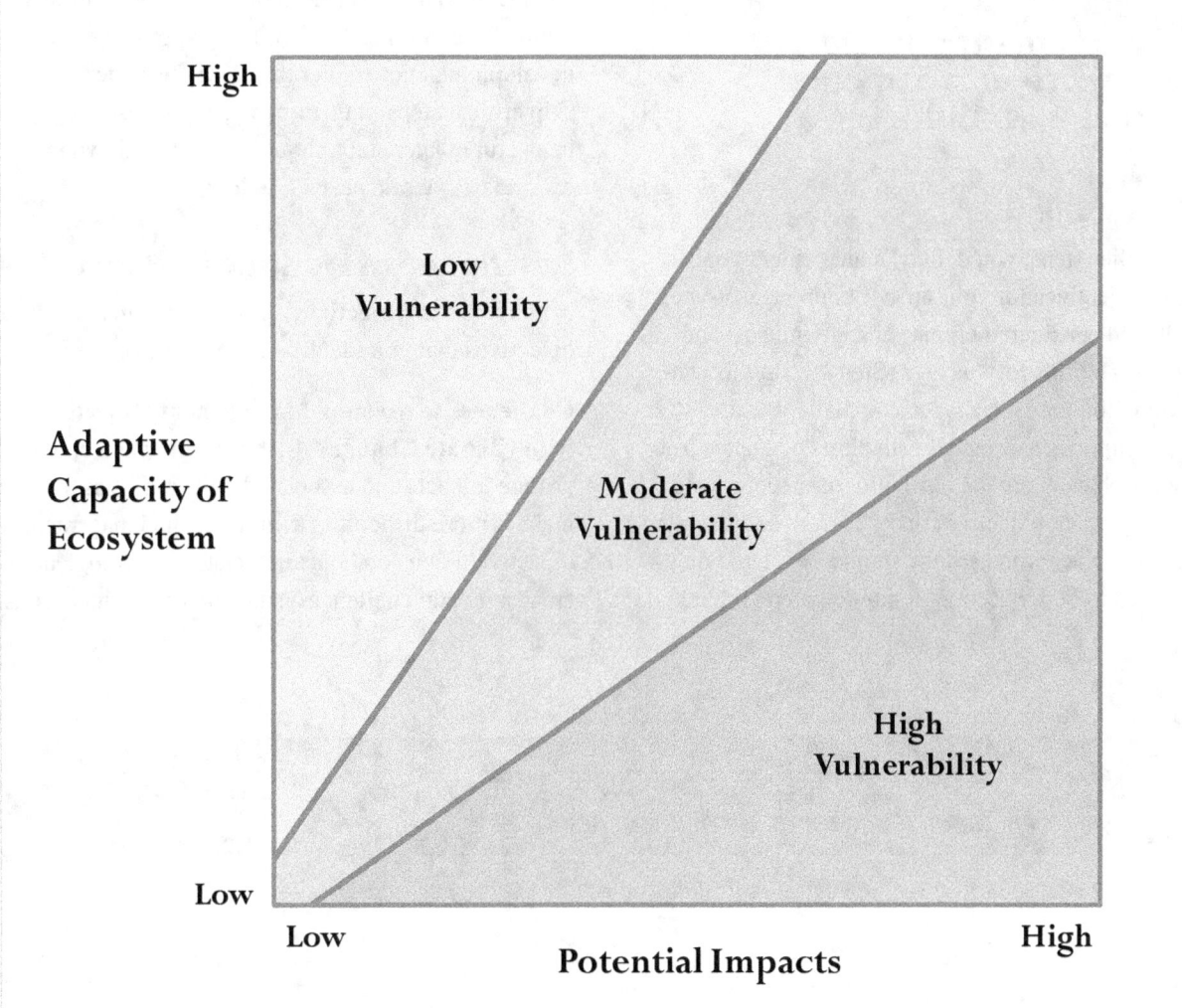

Figure 8.—The vulnerability determination considers an ecosystem's sensitivity to climatic changes, its exposure to those changes, and its capacity to adapt to those changes with minimal disruption (Glick et al. 2011, Levina and Tirpak 2006).

- **High** – Potential climate change impacts are expected to exceed the ability of the ecosystem to cope with impacts. Ecosystems may undergo changes that will disrupt important ecosystem functions and key environmental benefits.

- **Moderate** – Potential climate change impacts are expected to cause alterations to ecosystems, but ecosystems will be able to cope with some impacts.

- **Low** – Ecosystems are expected to readily cope with potential climate change impacts. It is not anticipated that climate change will have substantial negative effects on important ecosystem functions and environmental benefits.

45

Step #3: EVALUATE management objectives, given projected impacts and vulnerabilities.

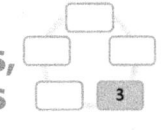

About this Step

In earlier steps, you defined management goals and objectives for your area of interest (Step #1) and considered climate change impacts and vulnerabilities for this area (Step #2). In this step, you will identify management challenges and opportunities associated with climate change. You will also evaluate the feasibility of meeting your management objectives under current management and consider whether they should be altered or refined to better account for anticipated climate

change impacts. It is inevitable that discussion will jump ahead at times to identifying approaches or developing tactics; rather than lose these ideas or skip critical steps in the process, any ideas that will be useful in later steps should be written down to revisit later (Worksheet #3, below).

Description of Worksheet #3 Items

Management Objectives – Insert the management objectives that you identified in Step #1 (column 5).

Challenges to Meeting Management Objective with Climate Change – List ways in which climate change impacts and associated vulnerabilities may make it more difficult to achieve each management objective. Focus on concerns related to ecological or environmental challenges, since other considerations

Worksheet #3. Full-size worksheets are located in the back of this document.

Management Objective (from Worksheet #1, column 5)	Challenges to Meeting Management Objective with Climate Change	Opportunities for Meeting Management Objective with Climate Change	Feasibility of Meeting Objective under Current Management	Other Considerations

(e.g., financial, social) will be included later in this step. For example, warmer temperatures and drier conditions may limit regeneration of a desired species and make it more challenging to maintain that species into the future.

Opportunities for Meeting Management Objective with Climate Change – List ways in which climate change impacts and associated vulnerabilities may make it easier to achieve each management objective or create new management opportunities. Focus on concerns related to ecological or environmental challenges, since other considerations (e.g., financial, social) will be included later in this step. For example, increases in small- and medium-scale disturbance may help increase structural heterogeneity within a stand or landscape.

Feasibility of Meeting Management Objective under Current Management – Consider the challenges and opportunities for managing under climate change in order to evaluate the feasibility of meeting your management objectives using *current* management strategies and actions. This feasibility determination is based primarily on ecological factors; other considerations are included in the next entry. Feasibility can be determined for individual or multiple time frames.

- **High** – Existing management options can be used to overcome the challenges for meeting management objectives under climate change. Opportunities likely outweigh challenges.

  Moderate – Some challenges to meeting management objectives under climate change have been identified, but these challenges can likely be overcome using existing management options. Additional resources or enhanced efforts may be necessary to counteract key challenges or promote new opportunities.

- **Low** – Existing management options may not be sufficient to overcome challenges to meeting management objectives under climate change. Additional resources or enhanced efforts will be necessary to counteract key challenges or promote new opportunities.

Other Considerations – List any social, financial, administrative, or other factors that would be part of your decision to pursue your management objectives. You may also want to note reasons why you would continue pursuing a management objective with low feasibility, such as requirement by law, a high social value, or likelihood of meeting management objectives across a broader geographic area (i.e., an area with low feasibility may still have the highest likelihood of success when compared to other areas with low feasibility).

 ## Slow Down to Consider...

It is very important to recognize that climate change may make management goals and objectives more difficult to achieve in the future (Joyce et al. 2008, 2009; Millar et al. 2007) and that there may be times in which they need to be altered or refined to better account for anticipated climate change impacts.

After completing the worksheet in Step #3, you should have a much better idea about whether your management objectives are feasible, given the current management options that are available to you. You've also identified social, economic, or other considerations that may affect your decision to pursue certain management objectives.

If you have high or extremely high feasibility of meeting all of your management objectives and these objectives are still sound, given projected climate change impacts, proceed to Step #4.

If some or all of the management objectives that you've identified have moderate feasibility or lower, or if they no longer seem sensible under climate change (e.g., managing a species that is very likely to experience a severe decline), you may want to consider whether you want to continue pursuing your management objectives, as well as your broader management goals. Additionally, because management objectives were assessed under the assumption of current management practices, there may be new or different management actions that could be used to meet your goals and objectives.

Are you going to continue with the management objectives that you have identified?

If yes:

- You may choose to provide documentation for why you will continue under Other Considerations in Step #3 (column 5).
- Proceed to Step #4 to explore Adaptation Strategies and Approaches.
- Note that you may choose to return to Step #1 and alter your management objectives at any point.

If no:

- Return to Step #1 to alter your management objectives or to develop new goals or objectives. Use the information that you have gathered up to this point to create goals and objectives that are more likely to succeed, given projected impacts from climate change.

Step #4: IDENTIFY adaptation approaches and tactics for implementation.

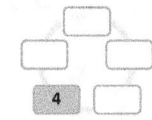

About this Step

New or modified management practices may be needed to address the challenges to ecosystem management brought about by climate change. In this step, you will actively brainstorm management approaches for climate change adaptation that address the challenges identified in the previous step. Then, drawing from the approaches, you will develop and evaluate tactics that describe the type and timing of management activities needed to achieve your management objectives and overcome the challenges you've identified (Worksheet #4, below).

This step is designed to use Chapter 2: Adaptation Strategies and Approaches, where a number of adaptation approaches have been summarized and described for forest types and conditions common in northern Wisconsin. These approaches were identified based on a number of expert reviews and may provide valuable options to guide management under climate change. However, because there may be adaptation approaches that have not been identified in the Adaptation Strategies and Approaches chapter, you may also want to develop additional approaches for meeting the management goals and objectives for your area of interest (Box 9).

Worksheet #4. Full-size worksheets are located in the back of this document.

Adaptation Approach	Tactic	Time Frames	Benefits	Drawbacks and Barriers	Practicability of Tactic	Recommend Tactic?

A
D
A
P
T
A
T
I
O
N

W
O
R
K
B
O
O
K

A
D
A
P
T
A
T
I
O
N

W
O
R
K
B
O
O
K

Box 9: Adaptation Approaches

The Adaptation Strategies and Approaches chapter of this document compiled approaches for forest adaptation from many resources and used expert feedback to refine the approaches for northern Wisconsin forest types. If you want to develop additional adaptation approaches, you may find it helpful to consult the following resources:

- *Strategies for Managing the Effects of Climate Change on Wildlife and Ecosystems*, prepared by the Heinz Center.
- *Biodiversity Management in the Face of Climate Change: a Review of 22 Years of Recommendations* by N.E. Heller and E.S. Zavaleta.
- *Managing for Multiple Resources under Climate Change: National Forests* by L. Joyce et al.
- *Forestry Adaptation and Mitigation in a Changing Climate: a Forest Resource Manager's Guide for the Northeastern United States* by J.S. Gunn et al.
- *Climate Change and Forests of the Future: Managing in the Face of Uncertainty* by C.I. Millar et al.
- *Adaptation to Climate Change in Forest Management* by D.L. Spittlehouse and R.B. Stewart.
- *Responding to Climate Change on National Forests: a Guidebook for Developing Adaptation Options* by D.L. Peterson et al.
- *Climate Project Screening Tool: an Aid for Climate Change Adaptation* by T.L. Morelli et al.
- *Adapting to Climate Change at Olympic National Forest and Olympic National Park* by J.E. Halofsky et al.

Description of Worksheet #4 Items

Adaptation Approach – Review the Adaptation Strategies and Approaches presented in this document (Chapter 2). Select any approaches that may be helpful in reaching your management goals and overcoming the challenges identified in Step #3. Include approaches that will help manage potential catastrophic events (pre- and post-disturbance), as well as any additional approaches that you devise that are not included in Chapter 2.

Tactics – Describe more specific actions that you can take in your area of interest to implement the adaptation approaches using your experience and expertise. You may have several tactics that can be used to implement a single approach, or one tactic that addresses multiple approaches. For example, an approach that favors existing species that are better-

suited to future conditions may include several tactics, including favoring future-adapted species that are present on site, modifying stand structure to increase natural regeneration of future-adapted species, and planting some areas using southern genotypes of species currently not found in the area of interest.

Time Frame(s) – List the approximate time frame(s) in which the tactics would be implemented. The nature of the action can help determine an appropriate time frame. Some actions may occur in the near term (i.e., next 2 years), while others may not occur for several decades or will occur only in certain situations (such as after a large disturbance). As a default, use the following categories to identify your relevant time frame(s), but allow the management objectives and ecosystems to dictate the most appropriate time frames:

- Immediate: 2 years or less
- Short-term: 2-10 years
- Medium-term: 10-30 years
- Long-term: 30 or more years

Benefits – For each tactic, list any benefits associated with using this tactic. For example, note if a tactic addresses your biggest challenge, addresses multiple challenges, or has a side benefit, such as improving overall ecosystem health.

Drawbacks and Barriers – For each tactic, list any drawbacks that may arise, such as negative ecosystem impacts, or any barriers to implementing the tactic, including legal, financial, infrastructural, social, or physical barriers.

Practicability of Tactic – Consider the benefits, drawbacks, and barriers associated with each tactic in order to determine the practicability of meeting your management goals and objectives using that tactic (Box 10). This determination is based on both ecological and non-ecological factors.

- **High** – The tactic is expected to be both effective and feasible. Benefits of the tactic clearly outweigh drawbacks and barriers.
- **Moderate** – There are drawbacks or barriers that could reduce the effectiveness or feasibility of the tactic. Some drawbacks or barriers may be overcome through the use of other adaptation tactics or management actions.

Box 10: What's Practicable?

An adaptation tactic is practicable if it is both effective (it will meet the desired intent) and feasible (it is capable of being implemented). Both of these characteristics increase the likelihood of success and are desirable in selected adaptation tactics.

- **Low** – The tactic does not appear to be effective or feasible. The drawbacks and barriers are insurmountable or the benefits are too small relative to the required effort. The tactic may need adjustment to be made more effective or feasible.

Recommend Tactic? – Consider the time frame, benefits, drawbacks, barriers, and practicability for each tactic and select the tactics that you recommend for consideration in future management decisions. Tactics that overcome or avoid challenges, have high practicability, or have major benefits should be favored. Box 11 contains additional considerations for evaluating the tactics.

For each tactic, determine whether you would recommend it for consideration in future management decisions:

- **Yes** – This tactic will likely be helpful in overcoming management challenges from climate change and meeting management objectives, and it should be considered in future management decisions. If needed, note any barriers that need to be overcome to use this tactic.
- **No** – This tactic is not helpful in overcoming management challenges or meeting management objectives, and it is not recommended for current consideration in future management activities.

It is important to reiterate that this workbook is intended to supplement existing decisionmaking processes, not replace them. In this step you are recommending further consideration of tactics in subsequent decisionmaking processes, but additional consideration does not mean that the tactics must be implemented or that the recommendations must supersede other considerations.

A
D
A
P
T
A
T
I
O
N

W
O
R
K
B
O
O
K

Box 11: Evaluating Tactics for Climate Change Adaptation

This Adaptation Workbook is designed to streamline the process of making decisions to help forest ecosystems adapt to climate change, but it is important to recognize that these types of decisions will always be complex and a number of variables need to be considered. For example:

- **Likelihood of Success** – Is an adaptation tactic likely to be effective in the future, given a variety of potential conditions? Will implementation of an adaptation tactic help achieve existing management objectives and goals?

- **Tradeoffs** – What are the potential tradeoffs associated with selecting and implementing a tactic? If you select an adaptation tactic, will there be negative consequences on other parts of the ecosystem or on other management actions? Can adverse impacts be avoided or mitigated?

- **Urgency** – Is there a need to implement an adaptation tactic in the near term? Will implementing a tactic now provide clear benefits in the future?

- **Cost** – Does an adaptation tactic have a high financial cost? Will implementing a tactic now prevent greater costs in the future?

- **Effort** - Is the adaptation tactic labor- or time-intensive? Will implementing a tactic now reduce the amount of work needed in the future?

The considerations above can be used to help select and prioritize approaches. In the short-term, you may want to emphasize tactics that can be described as "low-hanging fruit" because they are relatively easy to implement, have a high likelihood of success, and have few or no negative tradeoffs. At the same time, you can also identify barriers (such as cost, institutional structures, or social constraints) for tactics that you may want to implement in the future and work toward improving the feasibility of these actions for future efforts.

 **Slow Down to Consider...**

It is important to have a suite of management approaches that address potential challenges and help to meet management objectives. After completing the worksheet in Step #4, you should have a much better idea about whether the approaches and tactics that you selected will help address the management challenges that you identified. If the tactics that you've selected don't address all challenges or if many have low practicability, you may want to consider additional approaches or tactics before moving on to Step #5.

If you have decided that the identified challenges to meeting your management objectives cannot be overcome even after considering all possible tactics, you may want to alter or refine the tactics to be better aligned with the anticipated climate change impacts. Similarly, if there are substantial challenges to meeting your management objectives that may not be able to be overcome, you may also want to evaluate whether you will be able to achieve your broader management goals.

Are you going to continue with the adaptation tactics that have been selected?

If yes:

- For any tactic with moderate, low, or extremely low practicability, you may want to record the reasons that you are proceeding with that tactic in the last column of Step #4.
- Proceed to Step #5.
- Note that you can return to Step #1 and alter your management goals and objectives at any point.

If no:

- You may want to evaluate additional approaches and tactics before moving on to Step #5. You can reread the Adaptation Strategies and Approaches in Chapter 2, read other papers and documents on climate change adaptation, and consult with colleagues to identify other approaches that may be viable.
- If you were not able to identify approaches and tactics that could be used to meet your management objectives, you may want to return to step #1 to modify your objectives or to develop new goals or objectives. Use the information that you have gathered up to this point to create goals and objectives that are more likely to succeed, given projected impacts from climate change.

Step #5: MONITOR and evaluate effectiveness of implemented actions.

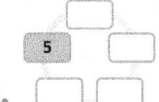

About this Step

Monitoring is critical for understanding what changes are occurring as a result of climate change as well as whether selected actions were effective in meeting management goals and adapting forests to future conditions. This step helps to identify metrics that will be used to monitor whether management goals are achieved in the future and to determine whether the recommended management tactics were effective. The outcome of this step is a list of realistic and feasible items that can be

monitored over time and used to help determine whether management should be altered in the future to account for new information and observations (Worksheet #5).

There are several types of monitoring, and many efforts are already underway to monitor some indicators in northern Wisconsin (Appendix 1). Most of these efforts are not designed to specifically monitor climate change, but they can still be useful in the context of climate change. Drawing upon and contributing to existing monitoring efforts when possible will help to detect changes that may not be detectable at smaller spatial scales and may also require fewer resources to implement. Consider what existing monitoring efforts are available and if they

Worksheet #5. Full-size worksheets are located in the back of this document.

Monitoring Items	Monitoring Metric(s)	Criteria for Evaluation	Monitoring Implementation

need to be modified to better monitor the results of your adaptation actions. Also consider what new monitoring items may be needed to evaluate whether you have met your management goals.

Description of Worksheet #5 Items

Monitoring Items – Identify monitoring items that will be used to evaluate whether you have achieved your management objectives and goals, or whether you have achieved a milestone that indicates that you are working toward your goal. When possible, select monitoring items that will also help you to understand whether the adaptation tactics recommended in the previous step were effective in working toward your management goals under climate change.

Monitoring Metric – Identify a metric for each monitoring item that can be used to evaluate your monitoring item. For example, if a monitoring item is to determine whether the conifer component within a stand was successfully increased through management activities, a metric could be the basal area of hemlock in the stand in 5, 10, and 20 years.

Criteria for Evaluation – Identify a criterion (e.g., condition or threshold) to evaluate whether the management goal was met or the tactic was successfully implemented. In the hemlock example above, a specific basal area value could be given to determine whether the conifer component was effectively increased.

Monitoring Implementation – Describe when information on the metric will be gathered and how the information will be collected (e.g., forest inventory data). The use of existing monitoring efforts is encouraged and some examples of ongoing efforts in northern Wisconsin are listed in Appendix 1. In completing this item, focus on creating a monitoring implementation plan that can be realistically carried out for the necessary period of time.

ADAPTATION WORKBOOK

Next Steps

By using this Adaptation Workbook, you have considered the effects of climate change on your area of interest. You have also identified management tactics and monitoring efforts to help you meet your management objectives under a changing climate. Now that you have completed this very important step toward improving the ability of your area of interest to adapt to the anticipated effects of climate change, you can work to integrate the information from the workbook, especially Step #4 and Step #5, into existing management plans and decisionmaking processes.

As you work toward this integration, it is important to keep in mind that the tactics you developed by completing the Adaptation Workbook have been recommended for further consideration (Step #4). Taking this step does not necessarily mean, however, that the tactics must be implemented or that the recommendations must supersede other considerations. The workbook was designed to lead you through a process for considering climate change, and it is up to you and your organization to determine the ways in which you will use the information and ideas you have developed.

Finally, the workbook is designed as part of an adaptive management process, which by definition needs to be able to incorporate new information as it becomes available. When developing a plan to implement your adaptation tactics and then monitor the results, also make plans to revisit this workbook as often as necessary to evaluate whether any changes are needed. Consult with experts whenever possible to gather new information and further refine your management decisions. As new information becomes available through scientific research, monitoring activities, or other avenues, use that information to consider how it may change your expectations regarding future conditions and whether it is appropriate to adjust your management or monitoring to better help the systems adapt to a changing climate.

CHAPTER 4: ADAPTATION ILLUSTRATIONS

Maria Janowiak, Patricia Butler, Chris Swanston, Matt St. Pierre, and Linda Parker

In this chapter, we demonstrate how the Adaptation Workbook (Chapter 3) can be used with the Adaptation Strategies and Approaches (Chapter 2) to develop adaptation tactics for two real-world management issues. The two illustrations in this chapter are intended to provide helpful tips to managers completing the Adaptation Workbook, as well as to show how the anticipated impacts of climate change can be addressed during land management planning and activities. Additionally, connections between the resources included in this document and other sources of information, such as vulnerability assessments, are highlighted in this chapter.

About the Illustrations

We have prepared the adaptation illustrations in this chapter based on information provided from land managers on the Chequamegon-Nicolet National Forest (CNNF). In fall 2010, we collaborated with two teams, each containing four to five land managers of different specialties (such as silviculture, wildlife, and hydrology). We worked with each team to define a unique management issue of interest and held a series of meetings where the managers completed the Adaptation Workbook. The teams considered their management issue at two very different scales of management; one team worked on a forest-wide issue at the programmatic level, and the other team worked on a single project area. The illustrations below contain our interpretation of the ideas, issues, and responses

that the two teams developed. In this chapter, we show how the Adaptation Workbook was used to consider climate change in forest management as a way to provide guidance for those learning to use the Adaptation Workbook (Box 12).

Box 12: Using the Illustrations

The Illustrations can:

- Provide an example of how the Adaptation Workbook was applied to a "real world" situation where climate change was considered with respect to ongoing management activities.

- Help managers understand what type of information is used in each step of the Adaptation Workbook and get a sense for the time and effort needed to complete each step.

- Present a short "case study" of how land managers are approaching adaptation in northern Wisconsin.

The Illustrations do not:

- Show the only possible ways that the Adaptation Workbook could be used to address resource management issues.

- Provide examples that will necessarily be implemented on the ground.

During the original testing of the Adaptation Workbook, the workbook contained six steps. Following the two teams' completion of the workbook, we sought the teams' comments about the efficiency and utility of the process, and how well it supported natural thought processes. Input from these discussions was later used to revise the process and condense it into five steps. For purposes of demonstration, and with consultation with the two teams, we present these illustrations as if they had originally occurred in five steps.

Illustration 1: Paper Birch Forest

For this first illustration, a team from the CNNF's Great Divide District, located in Ashland, Bayfield, and Sawyer counties, identified a project area containing paper birch that was under consideration for management within the next 5 years. In using the Adaptation Workbook, the team selected for evaluation a small subset of the project area focusing on management of early-successional aspen and paper birch because these forest types have high vulnerability to climate change and high frequency across the District. The Adaptation Workbook results for the paper birch forest type are described in this section and more information can be found in Appendix 4.

The paper birch forest type, which is dominated by paper birch and a number of boreal tree species, covers approximately 2 percent of the CNNF (Swanston et al. 2011), much of which is managed as early-successional forest for a variety of forest management and recreation opportunities (CNNF 2004). Paper birch has been identified as a tree species that is likely to experience substantial declines as a result of climate change effects and subsequent declines in suitable habitat (Swanston et al. 2011). As a result, the ability to maintain the paper birch forest type over the long term is a concern to landowners in northern Wisconsin, including the CNNF.

Step #1: DEFINE area of interest, management goals and objectives, and time frames.

The first step of the Adaptation Workbook helps define the scope for this exercise, which will be used in subsequent steps.

- The illustration team defined the **area of interest** as the paper birch forest type within a future management project with an early-successional emphasis (Table 4).

- This area of interest also defined the **location** (the project area) and **forest type** (paper birch).

- The team identified two **management goals** and several **management objectives** for the area based upon the CNNF's Forest Plan (CNNF 2004). When possible, the mature paper birch stands will be naturally regenerated to maintain paper birch forest on the landscape; in other areas that are less suitable for paper birch regeneration, a transition to white pine forest will be made to increase diversity.

- When defining **time frames**, the team identified the implementation of management as occurring in the immediate future (within 2 years). The team had some difficulty identifying a longer-term time frame, but eventually decided to use the end of the next rotation to evaluate whether management goals were realized. While defining time frames was challenging in this step, the team found the time frames helpful in completing later steps.

Table 4.—Description of the area of interest, management goals and objectives, and time frames completed in Step #1 of the Adaptation Workbook.

Area of Interest	Location	Forest Type(s)	Management Goals	Management Objectives	Time Frames
Early-successional paper birch forest within the defined project area (in Management Area 1B: Early Successional Aspen, Mixed Aspen-Conifer, and Conifer) *More information on this Management Area is available on pages 3-4 and 3-5 of the 2004 Land and Resource Management Plan (CNNF 2004).*	Paper birch stands within the project area	Paper birch	1) Retain paper birch forest on the landscape.	1) Regenerate the existing mature paper birch to retain it on the landscape when desirable.	Implementation is expected in the immediate future (2 years or less). Many management goals will be realized in the long term as paper birch is regenerated. The end of the next rotation is in ~60 years.
			2) Increase species and structural diversity.	2) Regenerate or underplant white pine among the natural paper birch when: (1) opportunities exist to improve stand diversity, (2) paper birch regeneration isn't possible, or (3) site scarification is not possible or desired.	

Step #2: ASSESS climate change impacts and vulnerabilities for the area of interest.

To evaluate the potential effects of climate change and others stressors on the area of interest, the illustration team examined a number of broad-scale impacts and vulnerabilities (see Table 3 for a complete list of broad-scale impacts and vulnerabilities by forest type) that were drawn from the *Ecosystem Vulnerability Assessment and Synthesis* (Swanston et al. 2011).

- The items in the list of **broad-scale impacts and vulnerabilities** were reviewed, and the team discussed how each of the impacts and vulnerabilities may or may not affect the area of interest that was defined in Step #1. Questions like "Is this impact important to paper birch?" or "Is this site more or less vulnerable than average?" were helpful in focusing discussion on what was most important in the area of interest.

- While the illustration team characterized potential **climate change impacts and vulnerabilities for the area of interest** using the broad-scale impacts and vulnerabilities from the *Ecosystem Vulnerability Assessment and Synthesis* (Swanston et al. 2011), team members would also have examined reports and peer-reviewed papers for additional impacts and vulnerabilities where time allowed.

- Nearly all of the impacts and vulnerabilities from the *Ecosystem Vulnerability Assessment and Synthesis* (Swanston et al. 2011) were applicable to the area of interest in some form, and about half of them were modified to better reflect those for the area of interest. For these impacts and vulnerabilities, the illustration team added text to better describe how the area of interest would be affected by climate change (Table 5). For example, the team noted that because the paper birch stands in the project area were relatively old (>60 years), the existing paper birch were more susceptible to the insects, diseases, and other disturbances that are expected to increase in the future, as well as to interactions among impacts.

- After considering climate change impacts and vulnerabilities, the team made a **vulnerability determination** that the area of interest had a "high" level of vulnerability. Current challenges to regenerating paper birch combined with projected decreases in habitat suitability and increases in a number of stressors suggested that substantial impacts to the area of interest may occur in the long term and that there may be a limited ability to buffer these impacts.

Table 5.—A portion of the assessment of impacts and vulnerabilities completed in Step #2 of the Adaptation Workbook. The items listed under "broad-scale impacts and vulnerabilities" were derived from the list in Table 3.

Broad-scale Impacts and Vulnerabilities	Climate Change Impacts and Vulnerabilities for the Area of Interest	Vulnerability Determination
	How might broad-scale impacts and vulnerabilities be affected by conditions in <u>your area of interest</u>? • Landscape pattern • Site location, such as topographic position or proximity to water features • Soil characteristics • Management history or current management plans • Species or structural composition • Presence of or susceptibility to pests, disease, or nonnative species that may become more problematic under future climate conditions • Other....	
Warmer temperatures	Warmer temperatures	High
Longer growing seasons	Longer growing seasons	
Altered precipitation regimes	Altered precipitation regimes	
Drier soils during summer	Warmer and drier conditions may be a substantial challenge because paper birch is on the edge of its range.	
Projected reduction in habitat suitability for many common tree species	Many tree species are projected to have reduced habitat suitability, including paper birch, aspen, balsam fir, and other common associates. Pine species are somewhat less vulnerable. Oak species may be favored.	
Increased fire and wind disturbance	Fire suppression reduces regeneration by preventing the development of required site and seedbed conditions. Regeneration of paper birch is difficult because fire cannot be used in most stands; other site preparation methods can also be constrained by visual concerns, topography, and other factors. Hotter, drier conditions may increase probability of natural fire, but it is unlikely to occur at the desired time and place.	
Increased threats from insects, diseases, and invasive plants	Because paper birch in the area of interest is over-mature, it is more susceptible to all stressors (e.g., drought, insects) and their interactions.	

Step #3: EVALUATE management objectives given projected impacts and vulnerabilities.

In this step, the illustration team evaluated climate change-related management challenges and opportunities for the area of interest.

- Many **challenges** were identified for the first objective, which focused on maintaining paper birch (Table 6). The team discussed how natural regeneration of paper birch is often difficult because heavy scarification or prescribed burning is needed to prepare a seedbed of bare mineral soil and it can be very difficult to achieve the requisite conditions. Many stands are not well-suited to the relatively high level of site preparation that is needed because of topography, accessibility, or concerns about visual impacts.

- While drier conditions and increased occurrence of wildfire could provide management **opportunities** for paper birch under certain conditions, in most cases it was expected that regeneration would become even more challenging in the future as climate change amplifies existing stressors and management challenges.

- Several management challenges and opportunities were identified for the second objective focused on species diversification using white pine because of uncertainty about future conditions.

- Given the challenges of regenerating paper birch (the first management objective), the team rated the **feasibility of meeting objective under current management** as "moderate" in the short term. In contrast, the team identified the second management objective focusing on enhancing the white pine component in selected areas as having higher feasibility than paper birch regeneration.

- The illustration team thought that the feasibility of meeting the objectives depended upon the time frame being evaluated. As a result, the team rated feasibility of meeting management objectives for both short- and long-term time frames.

- The team also identified **other considerations** which could affect its ability to achieve management objectives, including institutional, economic, social, and other non-biological constraints.

Slow down to consider...

This part of the process can be easily overlooked, but it provides a critical time to step back and verify that the management objectives and goals identified earlier are still appropriate and attainable.

- In Step #3, many challenges were identified for maintenance of paper birch, in both the short term and the long term, and long-term feasibility was rated as "low". The team was aware of these challenges, and thought that it was very important to proceed with the existing management goal and work to sustain paper birch on the site.

- The team recognized paper birch as an important component in northern Wisconsin and that its maintenance was consistent with CNNF plans and guidelines. More importantly, because the stands in the area of interest are mature, the team noted that there is currently a small window of time available for regenerating paper birch in these stands. If no action were taken within the next few years, the paper birch in these stands would begin to die, and the forests would succeed to a later-successional forest type no longer suitable for paper birch regeneration. Lastly, because it was predicted that paper birch regeneration would become even more difficult in the future, the team observed that an opportunity existed to regenerate and establish paper birch soon (before climate change impacts increase further) in order to maintain it on the landscape for as long as possible.

- The team identified potential management challenges as well as opportunities for diversifying stands with white pine, and long-term feasibility of this management objective was ranked as "moderate". Because feasibility was rated as "high" in the short term and "moderate" in the long term, the team

Table 6.—A portion of the evaluation of management objectives completed in Step #3 of the Adaptation Workbook.

Management Objective (from Step #1)	Challenges to Meeting Management Objective with Climate Change	Opportunities for Meeting Management Objective with Climate Change	Feasibility of Meeting Objective under Current Management	Other Considerations
Regenerate the existing mature paper birch to retain it on the landscape when desirable.	Warmer temperatures and drier conditions will make it increasingly difficult to regenerate paper birch. There is potential for more rapid decline of the species due to the northward shift in range and the expected increase in stressors. Opportunities for prescribed burns may become less available if fire danger is elevated, making site preparation more difficult to achieve.	Increased wildfire occurrence may benefit paper birch regeneration if fire occurs under the right conditions. If sites become drier, species that compete with paper birch regeneration may be reduced on some sites. Beyond this area of interest, some hardwood stands may become more conducive to paper birch management in the future as site conditions change.	Short-term: Moderate Long-term: Low	Native American tribes are interested in maintaining paper birch bark sources for baskets, canoes, and other uses. Social resistance to prescribed burning for site preparation may increase in the future, especially if wildfire occurrence increases. Additional resources and support may be needed to perpetuate the species in the future; it is unknown whether these will be available at that time.
Regenerate or underplant white pine among the natural paper birch when: (1) opportunities exist to improve stand diversity, (2) paper birch regeneration is not possible, or (3) site scarification is not possible or desired.	Regeneration of white pine (a key species identified for increasing diversity) may become more difficult due to deer browse, competition from raspberry and other plants, dry site conditions, and insect and disease outbreaks. Premature losses of the shelterwood overstory from wind disturbance may increase white pine susceptibility to white pine weevil and other pests. The white pine stock that is planted now may not be adapted to future conditions.	Regeneration of white pine may become easier if site conditions become more favorable for white pine and less favorable for competition. It may be better to regenerate white pine now because conditions in the future (50+ years) are uncertain and may be less favorable. Beyond this area of interest, some hardwood stands may become more conducive to paper birch management in the future as site conditions change.	Short-term: High Long-term: Moderate	When converting paper birch to white pine, managers need to consider the Forest Plan guidelines on species composition. White pine seedlings need to be protected from deer browse for 5-10 years after planting. When available, opportunities should be considered to diversify stands with species that may be favored in the future, such as oak species.

maintained the identified goals and objectives moving forward. Similar to the other goal, the team identified an opportunity to establish white pine in the area of interest in the near future while conditions are known to be favorable.

Step #4: IDENTIFY adaptation approaches and tactics for implementation.

To complete this step, the illustration team began by examining the Adaptation Strategies and Approaches.

- The illustration team worked through the Adaptation Strategies and Approaches chapter by discussing each **adaptation approach** individually in order. While this process took a long time, it seemed to be more efficient in the long run than taking a more "scattershot" approach. For each approach, the team discussed whether it was applicable to the area of interest and, if so, what tactics might be used to apply the approach. When an adaptation approach seemed to address management objectives and climate change challenges, the team selected it.

- For each selected approach, the team described one or more **adaptation tactics** that could be used to implement the selected approach (Table 7). Overall, the team selected 25 of the 41 adaptation approaches. Approximately 30 tactics were developed, many of which fit under more than one adaptation approach.

- Many of the tactics that were identified were consistent with existing management plans and policies. For example, the CNNF already has guidelines in place for retaining underrepresented tree species in areas that are harvested. This action directly relates to one of the adaptation approaches (Maintain and restore diversity of native tree species).

- Several new tactics were also identified as ways to increase the ability of the area to adapt to new issues arising from climate change. For example, one adaptation tactic that was evaluated was to use white pine planting stock from a broader geographic area, such as southwestern Wisconsin. While current management guidelines on the CNNF recommend using seeds and seedlings from known sources and from within the climatic zone in which the planting will occur, the team recommended this tactic as one that could be considered for implementation.

- The majority of the tactics that were developed had immediate time frames because the team believed that the tactics could be implemented in the next few years along with other planned management activities.

- For each tactic, the team identified **benefits** and **drawbacks and barriers** associated with each approach. Then the team rated the **practicability** of each tactic.

- The illustration team weighed all of these considerations and selected several tactics to **recommend**. After completing the Workbook, the team will further evaluate these recommended tactics to determine whether or how these tactics will be applied.

Slow down to consider...

The illustration team identified many adaptation approaches and tactics that helped to meet their management goals and objectives and addressed the challenges that were identified in earlier steps. Therefore, they were comfortable recommending the adaptation tactics proposed for further consideration.

Table 7.—Selected adaptation tactics that were developed and evaluated in Step #4 of the Adaptation Workbook.

Adaptation Approach	Tactic	Time Frames	Benefits	Drawbacks and Barriers	Practicability of Tactic	Recommend Tactic?
Maintain or improve the ability of forests to resist pests and pathogens.	Treat selected over-mature paper birch stands with a shelterwood harvest followed by prescribed burning or mechanical site preparation. Prioritize the stands to be treated using a field check of site conditions.	Immediate (2 years or less)	Younger birch trees tend to be less vulnerable and more resilient to stressors. Addresses multiple challenges. Regenerating paper birch helps meet goals and objectives set out in the Land and Resource Management Plan.	Regeneration is not guaranteed after treatment. It is uncertain whether this approach reduces paper birch's long-term vulnerability to climate change. Success is often dependent upon site conditions.	Short-term: Moderate Long-term: Low	Yes
	On sites with an existing white pine seed source or advanced regeneration, treat selected over-mature paper birch stands with a shelterwood harvest and scarify for white pine. Underplant white pine to augment advanced regeneration if needed. Retain the overstory.	Immediate (2 years or less)	Maintains a desired forest type in stands where paper birch regeneration is not possible or desired.		High	Yes
	Adjust rotation age lengths to achieve age class distribution goals in the Land and Resource Management Plan.	Long-term (30 or more years)	Addresses the current situation, where nearly all stands are at the end of rotation age (older than 60 years). Creates diversity in age classes across the landscape, which may make stands less susceptible to some climate change impacts.	Because stands need to be regenerated very soon to maintain paper birch as a dominant species, this diversification must occur in the next rotation.	High	Yes

(Table 7 continued on next page)

Table 7 (continued).—Selected adaptation tactics that were developed and evaluated in Step #4 of the Adaptation Workbook.

Adaptation Approach	Tactic	Time Frames	Benefits	Drawbacks and Barriers	Practicability of Tactic	Recommend Tactic?
Use seeds, germplasm, and other genetic material from across a greater geographic range.	For stands where white pine is underplanted, purchase stock from inside and outside of the immediate area (e.g., from farther south, east, or west). Keep records of what was used at different locations for tracking results over time.	Immediate to short-term (10 years or less)	May introduce genotypes that are better adapted to future conditions.	Sources suited to warmer and drier conditions are limited. Some genotypes may be less adapted to current or future conditions. Current guidelines recommend using stock from within the same climatic zone.	Moderate	Yes

Step #5: MONITOR and evaluate effectiveness of implemented actions.

In this step, the illustration team selected several items to help monitor whether the adaptation tactics were effective in helping to meet the management goals and objectives, as well as whether the management objectives were being reached.

- Given the emphasis on regeneration needed to achieve the management objectives for the area of interest, many of the monitoring items focused on whether desired species were successfully retained or regenerated (Table 8). Additional monitoring items sought to determine which management tactics enhanced regeneration and long-term survival of target species.

- The illustration team discussed monitoring of both implementation (i.e., whether an action was implemented) and effectiveness (i.e., whether an action achieved its desired objective), and worked to focus on monitoring effectiveness wherever possible.

- When possible, monitoring metrics, criteria, and implementation plans used or expanded upon existing monitoring activities.

- The team made a concerted effort to utilize existing efforts for monitoring, such as scheduled field visits and routine data collection. However, the team also made suggestions for increased monitoring efforts that balanced the monitoring needs with the institutional ability to monitor.

Table 8.—Selected monitoring items that were identified in Step #5 of the Adaptation Workbook.

Monitoring Items	Monitoring Metric(s)	Criteria for Evaluation	Monitoring Implementation
Management objective 1: Regenerate the existing mature paper birch to retain it on the landscape when desirable.	Acres treated Acres regenerated	Passes stocking survey	Monitor seedling success during 3rd- and 5th-year stocking survey. If a stand fails, implement follow-up activity and update monitoring.
Management objective 2: Regenerate or underplant white pine among the natural paper birch regeneration when opportunities exist to improve stand diversity, when paper birch regeneration is not possible, or when site scarification is not possible or desired.	Acres underplanted or regenerated	Passes stocking survey	Monitor seedling success during 1st- and 3rd-year stocking survey. If a stand fails, implement follow-up activity and update monitoring.
Tactic: For stands where white pine is underplanted, purchase stock from inside and outside of the immediate area (e.g., from farther south, east, or west). Keep records of what was used at different locations for tracking results over time.	Number of trees planted from different sources or locations Number of sources used Survival rate	Short-term: Passes 1st- and 3rd-year stocking survey Long-term: follow-up survival and condition survey	Short-term: Monitor seedling success (by source or location) during 1st- and 3rd-year stocking survey. If a stand fails, implement follow-up activity and update monitoring. Long-term: Coordinate with research for long-term evaluation of stock from alternative sources or locations.

Summary: Paper Birch Forest Illustration

While the illustration team rated the area of interest as having a "high" vulnerability to climate change-related impacts, it was able to identify several adaptation tactics to help improve paper birch's resilience and work to maintain the forest type over the next rotation. At the same time, several tactics, such as those focused on tree species diversification, create opportunities for a greater range of response in the future. Many of the tactics that were identified were similar to management actions that are commonly implemented in paper birch forests, such as the retention of under-represented tree species, prevention and control of nonnative invasive species, and use of prescribed fire for site preparation when possible. Such overlap between existing practices and adaptation actions suggests that many sustainable management actions that do not directly take climate change into account can still be immensely important for responding to climate change.

Additionally, the illustration team was able to identify adaptation tactics that are new options to consider for management. These tactics included prioritizing areas of strong paper birch regeneration as future refugia, expanding the geographic area from which seedlings are obtained, underplanting stands with low densities of oak species to increase diversity, and encouraging the growth of other species when efforts to regenerate paper birch have not been successful. For these "newer" tactics, the team emphasized a greater need for both monitoring and scientific assessment to determine whether the tactics are helping to reach management goals and maintain the integrity of these forest ecosystems.

Illustration 2: Spruce Grouse Habitat Management

For this second illustration, a team from the CNNF identified a forest-wide issue rather than a specific management area. The goal of using the workbook at this level was to identify broad-scale tactics to maintain or expand the representation of suitable habitat for spruce grouse on the CNNF. Spruce grouse was chosen as a focus for this illustration because it is a rare species in the region and many of the tree species that form its habitat are projected to experience substantial declines as a result of climate change.

Spruce grouse in Wisconsin are found almost exclusively in the northern two tiers of counties (Worland et al. 2009). Surveys on the CNNF indicate that the spruce grouse are primarily found in the lowland conifer forest type, which includes the peatlands dominated by black spruce and tamarack that are preferred habitat for spruce grouse (Worland et al. 2009). The lowland conifer forest type covers 14 percent (183,465 acres) of the CNNF (Swanston et al. 2011), and the type is not actively managed except where the action would benefit or maintain habitat for species of viability concern (CNNF 2004). Additionally, a juxtaposition of upland and lowland coniferous forests is important for the spruce grouse's lifecycle needs (Gregg et al. 2004, Worland et al. 2009). Many tree species found in both the lowland and upland spruce grouse habitat, including black spruce, white spruce, tamarack, and balsam fir, are projected to have large declines in suitable habitat due to climate change effects over the next 100 years (Swanston et al. 2011). As a result, spruce grouse habitat is vulnerable and of concern to CNNF managers. The illustration team completed the Adaptation Workbook for spruce grouse habitat in five different forest types simultaneously (lowland conifer, balsam fir, jack pine, spruce, and aspen), although many of the results presented in this section focus on the lowland conifer forest type.

Maps of spruce grouse suitable habitat.

Photo by Maria K. Janowiak, U.S. Forest Service and Northern Institute of Applied Climate Science

Step #1: DEFINE area of interest, management goals and objectives, and time frames.

The first step of the Adaptation Workbook helps define the scope of the management topic, which will be used in subsequent steps.

- The illustration team defined the **area of interest** as all habitat potentially suitable for spruce grouse across 1.5 million acres of the Chequamegon-Nicolet National Forest (Table 9). Consistent with the Conservation Assessment for the species (Gregg et al. 2004), optimal habitat for spruce grouse consists of black spruce-dominated lowland complexes and adjacent upland areas of young short-needled conifer habitat.

- Although suitable habitat for the species exists throughout the entire Great Lakes region, the **location** was limited to spruce grouse habitat within the CNNF in order to allow for the discussion to reach the tactical level. Currently, there are no formal partnerships to manage habitat across property boundaries.

- The team identified an overarching **management goal**, which emphasized the maintenance and improvement of spruce grouse habitat at the CNNF. More specific management objectives included maintaining existing habitat as well as creating new spruce grouse habitat when opportunities exist close to current habitat.

- The illustration team had some difficulty identifying the **time frames** but ultimately suggested that the long-term maintenance of optimal habitat would occur over the next 10-100 years and the establishment of new habitat would occur in the medium term (10-15 years). While defining time frames was challenging in this step, team members found the time frames helpful in completing later steps and adjusted the time frames slightly as they progressed through the Workbook.

Table 9.—Description of the area of interest, management goals and objectives, and time frames completed in Step #1 of the Adaptation Workbook.

Area of Interest	Location	Forest Type(s)	Management Goals	Management Objectives	Time Frames
Spruce grouse habitat on the CNNF	Stands that are of sufficient size. They may occur anywhere across the CNNF.	Lowland Conifer (dominated by black spruce)	Maintain and improve habitat for spruce grouse.	1) Maintain current spruce grouse habitat where it exists on the CNNF. 2) Create new habitat for spruce grouse as opportunities arise.	Long-term maintenance of optimal habitat: 10-100 years Establishment of new habitat: 10-15 years
	Stands that are adjacent to suitable lowland conifer complexes. They may occur anywhere across the CNNF.	Jack Pine (30 years or younger)			
		Balsam Fir (30 years or younger)			
		Spruce (30 years or younger)			
		Aspen (mixed aspen-spruce-fir)			

Step #2: ASSESS climate change impacts and vulnerabilities for the area of interest.

To evaluate the potential effects of climate change and other stressors on the area of interest, the illustration team reviewed many broad-scale impacts and vulnerabilities. Table 3 contains a list of broad-scale impacts and vulnerabilities by forest type.

- The team discussed how each of the items in the list of **broad-scale impacts and vulnerabilities** may or may not affect the area of interest defined in Step #1.

- Although we provided information on impacts and vulnerabilities from the *Ecosystem Vulnerability Assessment and Synthesis* (Swanston et al. 2011), this team also relied heavily on information from the *Climate Change Tree Atlas* (Prasad et al. 2007). The team also discussed how a watershed vulnerability assessment that was in development in the area could be used to provide additional information on impacts to forested wetlands in the future. Many other sources of information could be included in this step where appropriate, especially as new information becomes available.

- In the lowland conifer forest type, nearly all of the broad-scale climate change impacts and vulnerabilities were applicable to the area of interest. The team provided more information for about a third of these impacts and vulnerabilities to better describe how climate change impacts would be amplified or buffered in the areas of interest (Table 10). For example, the team noted that the substantial decline in black spruce habitat suitability that is projected for the end of the 21st century is a critical threat to stands dominated by black spruce. Additionally, the team added two site-specific vulnerabilities to the list (e.g., drying of peatlands may result in increased fire risk).

- After considering the area's vulnerability to climate change, the team made a **vulnerability determination** that the area of interest had a "high" level of vulnerability. Projected decreases in lowland conifer habitat suitability and increases in a number of stressors suggested that substantial impacts to the area of interest may occur in the short term and long term and that there may be a limited ability for management to buffer these impacts.

- The team also noted that across the northern Wisconsin landscape, species at the extreme southern edge of their range may be more vulnerable to climatic changes. The team did not list the area as having "extremely high" vulnerability because uncertainty about projected hydrological changes prevented them from declaring that the system will be severely disrupted and unable to provide key ecosystem benefits.

Table 10.—A portion of the assessment of impacts and vulnerabilities completed in Step #2 of the Adaptation Workbook for the lowland conifer forest type. The items listed under "broad-scale impacts and vulnerabilities" were derived from the list in Table 3.

Broad-scale Impacts and Vulnerabilities	Climate Change Impacts and Vulnerabilities for the Area of Interest	Vulnerability Determination
	How might broad-scale impacts and vulnerabilities be affected by conditions in <u>your area of interest?</u> • Landscape pattern • Site location, such as topographic position or proximity to water features • Soil characteristics • Management history or current management plans • Species or structural composition • Presence of or susceptibility to pests, disease, or nonnative species that may become more problematic under future climate conditions • Other….	
Warmer temperatures	Warmer temperatures	High
Longer growing seasons	Longer growing seasons	
Altered precipitation regimes	Altered precipitation regimes	
Drier soils during summer	Drier soils during summer; increased potential for drought	
Projected reduction in habitat suitability for many common tree species	Projected reduction in habitat. Many tree species are projected to have reduced habitat suitability, including black spruce, balsam fir, and tamarack. Tamarack may be more likely to persist on some sites despite drier conditions. The Climate Change Tree Atlas projects declines in black spruce habitat of up to 90 percent by the end of the 21st century.	
Decline of associated rare species	Decline of associated rare species	
Decline of associated wildlife species	Decline of associated wildlife species, including spruce grouse	
Potential reduction in the duration of soil saturation or ponding as a result of altered hydrology and precipitation patterns	Greater likelihood that sites will dry. Lowland conifer areas may receive reduced runoff during the growing season due to higher evapotranspiration losses and lower soil saturation in adjacent upland areas.	
Increased risk of fire occurrence in dried organic soils	Increased likelihood of fire. Peatland fires are currently uncommon. Patches of this type tend to be large and have fewer roads than other types. Therefore, fires that get established may be more extensive.	

Step #3: EVALUATE management objectives given projected impacts and vulnerabilities.

In this step, the illustration team evaluated climate change-related management challenges and opportunities for the area of interest.

- Many **challenges** were identified regarding the long-term maintenance of spruce grouse habitat (Table 11). Most of these challenges are direct results of increases in temperature and decreases in water availability. Other challenges, such as increased fire risk and mortality from combined stressors, are indirect effects of climate change that could result in catastrophic loss of this forest ecosystem.

- Some **opportunities** were identified where drier conditions are suitable for other short-needled conifer species.

- The team agreed that the feasibility of establishing optimal habitat over the next 15-30 years is high because some challenges to meeting management objectives under climate change can be overcome with existing management options. Feasibility beyond 30 years is uncertain and probably much lower.

- The team also listed **other considerations**, such as spruce grouse's status as a State Threatened species, that influence the ability and priority of committing resources to achieve the management objectives. The team also concluded that the likelihood of maintaining lowland complexes in the future is low to extremely low because the challenges to meeting management objectives are probably too great to overcome on a longer time scale.

Slow down to consider...

Although this part of the process is easy to overlook, it provides the opportunity to step back and verify that the management goals and objectives identified earlier are still appropriate.

- In Step #3, many challenges were identified for maintaining spruce grouse habitat, in both the short term and the long term, and long-term feasibility was rated as "extremely low". The team was aware of these challenges, and thought that there were several reasons to continue work to maintain and create spruce grouse habitat in the short term.

- The team recognized lowland conifer species as a critical habitat component for spruce grouse and many other boreal wildlife species and noted that its maintenance was consistent with CNNF plans and guidelines. Furthermore, the presence of spruce grouse on the Regional Forester Sensitive Species List emphasized the importance of maintaining or creating spruce grouse habitat. Opportunities to integrate spruce grouse habitat management with other programs or activities may allow easier and more effective landscape-scale planning.

- The team decided that the feasibility of creating new habitat in the short term was high. Despite the low feasibility of long-term maintenance of spruce grouse habitat under current management, the team noted the importance of maintaining habitat for this State Threatened species. Acknowledging the species' status prompted the team to move forward with the management objectives it had identified and look for adaptation tactics to improve the ability to maintain spruce grouse habitat into the future.

Table 11.—A portion of the assessment of challenges and opportunities to meeting management objectives completed for the lowland conifer forest type in Step #3 of the Adaptation Workbook.

Management Objective (from Step #1)	Challenges to Meeting Management Objective with Climate Change	Opportunities for Meeting Management Objective with Climate Change	Feasibility of Meeting Objective under Current Management	Other Considerations
1) Maintain current spruce grouse habitat where it exists on the CNNF. 2) Create new habitat for spruce grouse as opportunities arise.	Decreased regeneration of black spruce and balsam fir is likely. There is potential for high mortality of mature trees due to combined stress factors. Drier conditions may increase likelihood of invasion by nonnative species. Drier conditions from altered hydrology could result in black spruce trees that are less stunted, making them usable by spruce grouse for a shorter period of time. Tree density could become reduced below levels preferred by spruce grouse. Peat fire could result in "ponding," which would make the site unsuitable for spruce grouse. Encroachment of non-preferred tree species, such as white pine, could make a stand less suitable for spruce grouse.	Predominantly open bog/peatland systems may become suitable habitat in the future as they become drier. Lowland conifer systems could gradually become suitable for other short-needled conifers, such as jack pine.	Long-term maintenance of optimal habitat: High (short term) Low to extremely low (long term) Creation of new habitat: High	Spruce grouse is a State Threatened species and Regional Forester Sensitive Species. There are legal requirements as well as CNNF policy to maintain viability of this species. Some lowland conifer in spruce grouse habitat complexes are in areas where active habitat manipulation is discouraged. This habitat is a hotspot for numerous other rare elements (e.g., orchids). Seed availability for artificial regeneration of black spruce is limited. It is at the southern edge of its range, so there is no seed source farther south.

Step #4: IDENTIFY adaptation approaches and tactics for implementation.

Given the large spatial scale that was used for this illustration, the intent was to develop adaptation actions that are applicable to a range of specific sites and conditions, while allowing for flexibility in the management decisions made at those sites.

- Because spruce grouse habitat includes many forest types across a large area, the illustration team considered all five forest types when completing this step. This exercise allowed the team to take a broader view of the landscape and think about how adaptation actions applied differently to the various forest types.

- The illustration team worked through the list of adaptation strategies and approaches by discussing each **adaptation approach** individually in order. While this was a time-consuming process, it seemed to be more efficient overall than taking a "scattershot" approach. When an adaptation approach seemed to address management objectives and climate change challenges, the team selected it and recorded any adaptation tactics that came to mind. Listing all approaches and tactics, whether they are likely to be selected or not, created a record of ideas and provided an opportunity to reconsider them later.

- The list of selected approaches was then reviewed individually to develop **adaptation tactics** to describe why and how to implement the approaches in the area of interest. For the lowland conifer forest type, the team selected 6 of the 39 adaptation approaches (and no additional approaches were created). The team identified 30 tactics across all 5 forest types, approximately 11 of which were applicable to the lowland conifer forest type.

- The majority of the tactics that were selected had immediate and short-term time frames that employ resistance approaches to improve the forest's defenses against anticipated changes. Two approaches were identified that can increase resilience or facilitate adaptation in the medium term and long term.

- The team discussed the **benefits**, as well as the **drawbacks and barriers**, for each tactic (Table 12). The team observed that uncertainty can also be evaluated in the drawbacks and barriers; some things may be uncertain now, but will likely be better understood in the future. Next, the team rated the **practicability** of the tactic. An adaptation approach is practicable if it is both feasible (i.e., it will help meet management objectives) and capable of being implemented. Adaptation tactics that are consistent with the Forest Plan are rated as having higher practicability. The team identified a number of tactics that can be implemented easily because they support the guidelines in the Forest Plan. For example, a variety of actions could be taken to manage water flow at control points to benefit lowland conifer hydrology, including road decommissioning, culvert replacement, and installation of water control structures.

- The team also identified new tactics that would require more information or greater flexibility in the Forest Plan to implement. The team acknowledged that when considering the "worst case" scenario of future conditions, extreme actions that are beyond what can currently be implemented may need to be considered. For example, if severe conditions were to cause many peatland forests to dry, it may make sense to consider artificially supplying needed water to key locations and systems.

- The team discussed the **time frames** for implementing tactics. Many can be implemented immediately to increase resilience or resistance, but several are meant to be triage responses (e.g., if hydrologic conditions become too dry some years into the future).

- The illustration team weighed all of these considerations and selected a number of tactics to **recommend**. After completing the Workbook, the team will evaluate these recommended tactics further to determine whether or how these tactics will be applied.

Table 12.—A portion of the selected approaches and devised tactics completed in Step #4 of the Adaptation Workbook

Adaptation Approach	Tactic	Time Frames	Benefits	Drawbacks and Barriers	Practicability of Tactic	Recommend Tactic?
Maintain or restore hydrology.	In lowland conifer forest type: Identify roads or other control points that affect hydrology in spruce grouse peatland complexes. Prioritize based on spruce grouse habitat qualities.	Immediate (2 years or less)	Addresses most significant challenge and other challenges. Managing wetland drainage has co-benefits of managing for wetland ecosystems, and benefits wetland ecosystem function.	Effectiveness is uncertain. Water yield manipulation could have undesirable effects on other management objectives.	Moderate	Yes
	In lowland conifer forest type: Based on prioritized list (above item), manage flow at control points to benefit hydrology. Examples include: decommissioning roads, replacing culverts, constructing berms or water control structures, and managing adjacent vegetation.	Short-term (2-10 years)	Addresses most significant challenge and other challenges. Managing wetland drainage has co-benefit of managing for wetland ecosystems and benefits wetland ecosystem function.	Effectiveness is uncertain. Water yield manipulation could have undesirable effects on other management objectives.	Moderate	Yes
	In lowland conifer forest type: Artificially supply or drain water to maintain water levels in high-priority peatlands.	Long-term (30 years or more)		Potential for many negative tradeoffs	Low	No
Maintain and create habitat corridors through reforestation or restoration.	Across all forest types: As part of the proposed Spruce Grouse Habitat Assessment, evaluate current level of connectivity between suitable spruce grouse habitat complexes. Map or identify current or potential corridors.	Long-term (30 or more years)	May allow for spruce grouse to disperse to currently unoccupied habitat or habitat that becomes suitable in the future.	Corridors can also serve invasive species, predators, and other undesirable species. May not be consistent with other landscape management objectives. Is suitable only in limited areas.	Moderate	Yes

Slow down to consider...

The illustration team identified a number of adaptation approaches and tactics to help meet its management objectives and address the challenges that were identified in earlier steps. Despite having identified only a few approaches and strategies that help meet management objectives in the long term, the team has considered and developed responses to substantial challenges in maintaining spruce grouse habitat.

Step #5: MONITOR and evaluate effectiveness of implemented actions.

In this step, the illustration team selected several items to help monitor whether the adaptation tactics were effective in helping to meet the management objectives, as well as whether the management goals and objectives were being reached (Table 13).

- Given the landscape-level focus of the management objectives for the area of interest, many of the **monitoring items** addressed whether spruce grouse were utilizing areas of potential habitat, and whether landscape planning for the spruce grouse was being integrated into project planning and management.

- When possible, **monitoring metrics, criteria, and implementatio**n plans used or expanded upon existing monitoring activities, such as tree stocking surveys.

- Monitoring is important in gathering data about the greatest uncertainties for spruce grouse habitat management. For example, current models are not able to predict tree species' responses in cooler sites or in other microenvironments. Monitoring will be valuable in determining where important tree species are more likely to persist and which management actions are helping to adapt ecosystems to climatic changes.

- The illustration team discussed two types of monitoring. One type focused on whether an action was implemented, such as whether recommended tactics were incorporated into the design criteria of a project. Another type looked more directly at whether an action was effective in meeting its desired objective. For example, while this illustration focused on maintaining spruce grouse habitat, the team recognized that it is important to monitor actual birds to evaluate whether they are using the habitat.

Table 13.—A portion of the selected monitoring metrics completed in Step #5 of the Adaptation Workbook.

Monitoring Items	Monitoring Metric(s)	Criteria for Evaluation	Monitoring Implementation
In lowland conifer forest type: Identify roads or other control points that affect hydrology in spruce grouse peatland complexes. Prioritize based on spruce grouse habitat qualities.	Monitor implementation: determine if complete.	Were control points and roads in all priority spruce grouse habitat complexes identified?	Check after 5 years.
In lowland conifer forest type: Identify roads or other control points that affect hydrology in spruce grouse peatland complexes. Prioritize based on spruce grouse habitat qualities.	Monitor effectiveness: metrics to be determined after consultation with a hydrologist.	Criteria to be determined after consultation with a hydrologist	Implementation to be determined after consultation with a hydrologist.
Across all forest types: As part of the proposed Spruce Grouse Habitat Assessment, evaluate current level of connectivity between suitable spruce grouse habitat complexes. Map or identify current or potential corridors.	Monitor implementation: determine if complete.	Was a Spruce Grouse Habitat Assessment completed?	Check annually for completion.
Across all forest types: Connect large lowland conifer peatland complexes with suitable upland habitat to allow spruce grouse dispersal to currently unoccupied habitat. For example, convert key stands to short-needled conifer types to increase connectivity between spruce grouse habitat complexes.	Monitor implementation: determine whether activities to encourage or convert to short-needled conifer are occurring in identified "key stands".	Presence of activity in key stands to encourage or convert to short-needled conifer	Review prescriptions and other documents within the identified key stands.
	Monitor effectiveness: assess usage by spruce grouse in connector stands.	Occurrence of spruce grouse in key stands	Conduct spruce grouse surveys.

Summary: Spruce Grouse Habitat Management Illustration

This team identified five forest types that were important for spruce grouse habitat across the landscape, which centered on lowland conifer complexes. Although this illustration has emphasized the discussions on lowland conifer systems, all forest types were considered simultaneously because of their relationships with each other and how spruce grouse populations might find them desirable. The vulnerabilities, tactics, and monitoring items identified by the team in the Adaptation Workbook have the potential to be incorporated into future management direction on spruce grouse habitat on the CNNF. Additionally, several of the approaches and tactics identified during this process may be useful to managers considering actions to maintain spruce grouse habitat.

GLOSSARY

Adaptation
Adjustments, both planned and unplanned, in natural and human systems in response to climatic changes and subsequent effects. Ecosystem-based adaptation activities use a range of opportunities for sustainable management, conservation, and restoration.

Adaptive capacity
The ability of a species or ecosystem to accommodate or cope with climate change impacts with minimal disruption.

Adaptive management
A dynamic approach to forest management in which the effects of treatments and decisions are continually monitored and used, along with research results, to modify management on a continuing basis to ensure that objectives are being met.

Age class
An age interval into which trees are divided for classification or use. A distinct aggregation of trees may originate from a single natural event or regeneration activity.

Approach
Adaptation response that can be applied to a single ecosystem or forest type.

Assemblage
A dynamic group of species that form a community.

At-risk species
A species that has been determined to be vulnerable to observed or projected changes in climate or other stressors.

Biological legacy
Individual trees of a variety of species retained from harvest in order to maintain their presence on the landscape, provide a potential seed source for both species and genotypes that are expected to be better adapted to future conditions, and serve as future nurse logs for regeneration of some species.

Biomass
The mass of living organic matter (plant and animal) in an ecosystem. Biomass also refers to organic matter (living and dead) available on a renewable basis for use as a fuel; biomass includes trees and plants (both terrestrial and aquatic), agricultural crops and wastes, wood and wood wastes, forest and mill residues, animal wastes, livestock operation residues, and some municipal and industrial wastes.

Boreal forest
A forest found only at latitudes of between 50-55° and 65-70° in the Northern Hemisphere and adapted to cool temperatures and low rainfall (below 20 inches).

Catastrophic event
In the context of forest management, an event caused by natural forces that results in near or total mortality of a species, community, or unit of forest.

Climate
The statistical description of the weather in terms of the mean and variability of relevant quantities (usually temperature, precipitation, and wind) over periods of several decades (typically three decades). In a wider sense, the "climate" is the description of the state of the climate system.

Climate change

A change in the state of the climate that can be identified (e.g., by using statistical tests) by changes in the mean and/or the variability of its properties, and that persists for an extended period, typically decades or longer.

Climate projection

A projection of the response of the climate system to emission scenarios of greenhouse gases based upon simulations by climate models. Climate projections are distinguished from climate predictions in order to emphasize that climate projections depend upon the scenario used, which is based on assumptions concerning, for example, future socioeconomic and technological developments that may or may not be realized and are therefore subject to substantial uncertainty.

Coarse woody debris

Any piece(s) of dead woody material, including dead boles, limbs, and large root masses, that are on the ground in forest stands or in streams.

Community

An assemblage of plants and animals living together and occupying a given area.

Conservation

The management of a renewable natural resource with the objective of sustaining its productivity in perpetuity while providing for human use compatible with sustainability of the resource. In managed forests, sustainable use may include periodic cutting and removal of trees followed by regeneration.

Disturbance

Stresses and destructive agents such as invasive species, diseases, and fire; changes in climate and serious weather events such as hurricanes and ice storms; pollution of the air, water, and soil; real estate development of forest lands; and timber harvest. Some of these are caused by humans, in part or entirely; others are not.

Diversity

The variety and abundance of life forms, processes, functions, and structures of plants, animals, and other living organisms, including the relative complexity of species, communities, gene pools, and ecosystems at spatial scales that range from local through regional to global. There are commonly five levels of biodiversity: (a) genetic diversity, referring to the genetic variation within a species; (b) species diversity, referring to the variety of species in an area; (c) community or ecosystem diversity, referring to the variety of communities or ecosystems in an area; (d) landscape diversity, referring to the variety of ecosystems across a landscape; and (e) regional diversity, referring to the variety of species, communities, ecosystems, or landscapes within a specific geographic region.

Downscaling

A method for obtaining high-resolution climate or climate change information from relatively coarse-resolution global climate models. Downscaling involves examining the statistical relationship between past climate data and on-the-ground measurements.

Ecological function

The sum of physical conditions (e.g., depth of water and soil type) and ecological processes (e.g., nutrient cycling and sediment movement) that make up an ecosystem and, ultimately, habitats on which species depend. A loss of ecological function is the removal or disruption of an ecological process that produces a certain physical condition or the loss or damage to a physical condition.

Ecological processes

Processes fundamental to the functioning of a healthy and sustainable ecosystem, usually involving the transfer of energy and substances from one medium or trophic level to another (e.g., water flows and movement, nutrient cycling, sediment movement, and predator-prey relationships).

Ecological Province
Climatic subzones, controlled primarily by continental weather patterns such as length of dry season and duration of cold temperatures. Provinces are also characterized by similar soil orders and are evident as extensive areas of similar potential natural vegetation, such as those mapped by Kuchler (1964). Defined by the National Hierarchical Framework of Ecological Units, these ecological units are grouped under Ecoregions.

Ecoregion
A region showing a repetitive pattern of ecosystems associated with commonalities in soil and landform. Defined by the National Hierarchical Framework of Ecological Units, ecoregions are recognized by differences in global, continental, and regional climatic regimes and gross physiography.

Ecosystem
A system of living organisms interacting with each other and their physical environment. The boundaries of an ecosystem are somewhat arbitrary, depending on the focus of interest or study. Thus, the extent of an ecosystem may range from very small spatial scales to, ultimately, the entire Earth.

Ecosystem driver
Any natural or human-induced factor that directly or indirectly causes a change in an ecosystem. A direct driver unequivocally influences ecosystem processes and can therefore be identified and measured. An indirect driver operates more diffusely, often by altering one or more direct drivers, and its influence is established by understanding its effect on direct drivers.

Evapotranspiration
The sum of evaporation and plant transpiration from the Earth's land surface to the atmosphere.

Even-aged management
A forest management method in which all trees in an area are harvested at one time or in several cuttings over a short time to produce stands that are all the same age or nearly so.

Feasibility
A determination of the ability to meet a management objective, in an ecological context, based on the sum of management opportunities and challenges arising from projected climate changes.

Forest type
A classification of forest land based on the dominant species present, as well as associate species commonly occurring with the dominant species.

Fragmentation
A disruption of ecosystem or habitat connectivity, caused by human or natural disturbance, creating a mosaic of successional and developmental stages within or between forested tracts of varying patch size, isolation (distance between patches), and edge (cumulative length of patch edges).

Framework
A conceptual structure or process intended to support decisionmaking, planning, and implementation.

Fuelbreak
A physical barrier to the spread of fire, such as a road, bulldozer line, or body of water. A fuelbreak can also be defined as a change in composition and density of a forest at its edges to reduce woody fuel for fires.

Gene flow
The consequence of cross-fertilization between members of a species that results in the spread of alleles across and between populations.

Genetic diversity
Genetic variation within a species.

Germplasm
Within an individual or group, the collective hereditary materials that are the physical basis for inheritance, that is, the transmission of hereditary materials (genotype) to the next generation.

Genotype
An individual's hereditary (genetic) constitution or individual(s) characterized by a certain genetic constitution.

Goal
Broad statements, usually not quantifiable, that express a desired state or process to be achieved. Goals are often not attainable in the short term, and provide the context for more specific objectives.

Greenhouse gas
The gaseous constituents of the atmosphere, both natural and anthropogenic, that absorb and emit radiation at specific wavelengths within the spectrum of thermal infrared radiation emitted by the Earth's surface, the atmosphere, and clouds. Water vapor (H_2O), carbon dioxide (CO_2), nitrous oxide (N_2O), methane (CH_4), and ozone (O_3) are the primary greenhouse gases in Earth's atmosphere.

Growing season
The period in each year when the weather and temperature are right for plants and crops to grow.

Habitat corridor
A defined tract of land connecting two or more areas of similar management or habitat type that is protected from substantial disturbance and through which a species can travel to reach habitat suitable for reproduction and other life-sustaining need.

Illustration
A case study of applying adaptation tools and resources to a management activity in a particular location in order to develop experience and examples of such applications for the benefit of a larger group of potential users.

Impact
The direct and indirect consequences of climate change on systems, particularly those that would occur without adaptation.

Impact model
Simulations of impacts on trees, animals, and ecosystems. These models use global circulation model projections as inputs, and include additional inputs such as tree species, soil types, and life history traits of individual species.

Invasive species
Any species that is nonnative (or alien) to the ecosystem under consideration and whose introduction causes or is likely to cause damage, injury, or disruption to ecosystem processes or other species within that ecosystem.

Landscape
(1) A spatial mosaic of several ecosystems, landforms, and plant communities across a defined area irrespective of ownership or other artificial boundaries and repeated in similar form throughout. For the purpose of this document, landscapes are often defined as ecoregions within the political boundary of one or several states (e.g., the portion of Ecological Province 212 in Wisconsin). Ecological units are defined by general topography, geomorphic process, surficial geology, associations of soil families, and potential natural communities, patterns, and local climates. (2) The term 'landscape level' or 'landscape scale' refers to activities or concepts that operate in or encompass a spatial mosaic of ecosystems, landforms, and plant communities.

Microhabitat
The specific combination of habitat elements in the locations selected by an organism for specific purposes or events. Distinctive physical characteristics and the more specific and functional aspects of habitat and cover distinguish the microhabitats within an organism's habitat.

Migration
The movement of genes, individuals, or species from one population or geographic location to another. Tree migration is largely influenced by dispersal ability, landscape connectivity, and climatological and other factors.

Mitigation
In the context of climate change, a human intervention to reduce the sources or enhance the sinks of greenhouse gases. Examples include expanding forests and other "sinks" to remove greater amounts of carbon dioxide from the atmosphere, using fossil fuels more efficiently, switching to solar energy or wind power, and improving the insulation of buildings.

Monitoring
The collection of information over time, generally on a sample basis by measuring change in an indicator or variable, to determine the effects of resource management treatments in the long term.

Monoculture
A stand of a single species, generally even-aged.

Native species
An indigenous species that is normally found as part of a particular ecosystem, and that was present in a defined area prior to European settlement.

Objective
Concise, time-specific statements of measurable planned results that correspond to preestablished goals in achieving a desired outcome.

Pest
An organism that is undesirable or detrimental to the interests of humans. In the context of ecology, pests are usually insects capable of causing injury, damage, or mortality to trees.

Practicability
A determination of the ability to attain a management goal based on both the effectiveness and the feasibility of implementing an adaptation tactic.

Precipitation
Any or all forms of liquid or solid water particles that fall from the atmosphere and reach the Earth's surface, including drizzle, rain, snow, snow pellets, snow grains, ice crystals, ice pellets, and hail.

Process model
A model that relies on computer simulations based on mathematical representations of physical and biological processes that interact over space and time.

Productivity
The rate at which biomass is produced per unit area by any class of organisms, or the rate of energy utilization by organisms.

Provenance
The original geographic source of seed, pollen, or propagules.

Realignment
The process of tuning ecosystems or habitats to present and anticipated future conditions in such a way that they can respond adaptively to ongoing change.

Refugia
Locations and habitats that support populations of organisms that are limited to small fragments of their previous geographic range.

Regeneration
The vegetative (e.g., sprouting from clonal root structures and coppicing) or sexual regeneration of a plant species.

Reserve
Natural areas with little to no harvest activity, unless required to maintain the system, that do not exclude fire management or other natural disturbance processes.

Resilience
An adaptation option intended to accommodate some degree of change, but allow for a return to prior conditions after a disturbance, either naturally or through management.

Resistance
An adaptation option intended to improve the defenses of an ecosystem against anticipated changes or directly defend the forest against disturbance in order to maintain relatively unchanged conditions.

Response
(1) The behavior of an individual under the influence of environmental changes. (2) An adaptation option intended to accommodate change and enable ecosystems to adaptively respond to changing and new conditions (Millar et al. 2007).

Restoration
The process of returning ecosystems or habitats to their original structure and species composition.

Riparian
Related to, living, or located in conjunction with a wetland, on the bank of a river or stream, or at the edge of a lake or tidewater. The riparian community significantly influences, and is significantly influenced by, the neighboring body of water.

Risk
The chance of something happening that will have an impact on objectives, often specified in terms of an event or circumstance and the consequences that may flow from it. Measured in terms of the consequences of an event and their likelihoods, risk may have a positive or negative impact.

Rotation
The period between regeneration establishment and final cutting. Rotation may be based on such criteria as mean size, age, culmination of mean annual increment, attainment of particular minimum physical or value growth rate, and biological condition.

Scenario
A coherent, internally consistent, and plausible description of a possible future state of the world. It is not a forecast; rather, each scenario is one alternative image of how the future can unfold. A projection may serve as the raw material for a scenario, but scenarios often require additional information.

Seed zone
A designated area, usually with definite topographic bounds based on climatological, biological, and geographical criteria, containing trees with relatively uniform genetic (racial) composition as determined by progeny-testing various seed sources.

Sequestration
The capture of carbon dioxide (CO_2) for long-term storage to either mitigate or defer the accumulation of CO_2 in the atmosphere.

Site quality
The productive capacity of a site, typically expressed as volume production of a given species, usually determined by the prevailing soil type and condition, moisture regimes, and local climatic conditions.

Snowpack
Layers of accumulated snow that usually melts during warmer months.

Species distribution model
A model that uses statistical relationships to project change in the distributional range of a given species.

Stand
A contiguous group of trees sufficiently uniform in age class distribution, composition, and structure, and growing on a site of sufficiently uniform quality, to be a distinguishable unit.

Static ecosystem
An ecosystem that is relatively sensitive to disturbance, and that has little capacity to return to pre-disturbance conditions.

Strategy
A broad adaptation response that considers regionally specific ecological and managerial conditions.

Stressor
An agent, condition, change in condition, or other stimulus that causes stress to an organism.

Succession
The gradual supplanting of one community of plants by another. Succession is primary on sites that have not previously borne vegetation, and secondary after the whole or part of the original vegetation has been supplanted.

Suitable habitat
In the context of the Climate Change Tree Atlas (a species distribution model), the area- weighted importance value, or the product of tree species abundance and the number of cells with projected occupancy (Swanston et al. 2011).

Sustainable forest management
The stewardship and use of forests in a way, and at a rate, that maintains their biodiversity, productivity, regeneration capacity, vitality, and potential to fulfill, now and in the future, ecological, economic, and social functions at local, national, and global levels, and that does not cause damage to other ecosystems.

Tactic
A prescriptive action designed for individual site conditions and management objectives.

Triage
In the context of forest management, a systematic process to sort management situations into categories according to urgency, sensitivity, and capacity of available resources to achieve desired goals. Cases are rapidly assessed and divided into major categories that determine treatment priority.

Uncertainty
An expression of the degree to which a value (e.g., the future state of the climate system) is unknown. Uncertainty can result from insufficient knowledge, quantifiable errors in the data, ambiguously defined concepts or terminology, or uncertain projections of human behavior. Uncertainty can therefore be represented by quantitative measures, for example, a range of values calculated by various models, or by qualitative statements, for example, reflecting the judgment of a team of experts.

Vigor
The general metabolic activity of a tree. Vigor is indicated by the healthy and robust growth of a tree.

Vulnerability
The susceptibility of a system to the adverse effects of climate change. Vulnerability is a function of the magnitude of climatic change, the sensitivity of a system, and the ability of the system to adapt.

Vulnerability assessment
The process of identifying, quantifying, or prioritizing the vulnerabilities in a system.

Wildland-urban interface
Any area where man-made improvements are built close to, or within, natural terrain and flammable vegetation, and where high potential for wildland fire exists.

LITERATURE CITED

Abrams, M.D. 1992. **Fire and the development of oak forests.** BioScience. 42(5): 346-353.

Agee, J.K.; Bahro, B.; Finney, M.A.; Omi, P.N.; Sapsis, D.B.; Skinner, C.N.; van Wagtendonk, J.W.; Weatherspoon, CP. 2000. **The use of shaded fuelbreaks in landscape fire management.** Forest Ecology and Management. 127(1-3): 55-66.

Ayers, M.P.; Lombardero, M.J. 2000. **Assessing the consequences of global change for forest disturbance from herbivores and pathogens.** Science of the Total Environment. 262(3): 263-286.

Biringer, J.L. 2003. **Forest ecosystems threatened by climate change: promoting long-term forest resilience.** In: Hansen, L.J.; Biringer, J.L.; Hoffman, J.R., eds. Buying time: a user's manual for building resistance and resilience to climate change in natural systems. Geneva, Switzerland: World Wildlife Fund: 43-71.

Brandt, L.A.; Swanston, C.W.; Parker, L.R., Janowiak, M.K.; Birdsey, R.A.; Iverson, L.R.; Mladenoff, D.J.; Butler, P.R. In Review. **Climate change science applications and needs in forest ecosystem management: a workshop organized as part of Chequamegon-Nicolet National Forest Climate Change Response Framework.** U.S. Department of Agriculture, Forest Service, Northern Research Station.

Burns, R.M.; Honkala, B.H. 1990. **Silvics of North America: 1. conifers; 2. hardwoods.** Washington, DC: U.S. Department of Agriculture, Forest Service: 877 p.

Byers, E.; Norris, S. 2011. **Climate change vulnerability assessment of species of concern in West Virginia.** Project Report. Elkins, WV: West Virginia Division of Natural Resources. Available at http://wvdnr.gov/publications/PDFFiles/ClimateChangeVulnerability.pdf. (Accessed 8 September 2011).

Carey, J.H. 1993. *Pinus banksiana.* In: Fire Effects Information System, [Online]. Fort Collins, CO: U.S. Department of Agriculture, Forest Service, Rocky Mountain Research Station, Fire Sciences Laboratory. Available at http://www.fs.fed.us/database/feis/. (Accessed 25 December 2011).

Chequamegon-Nicolet National Forest. 2004. **2004 Land and resource management plan.** Milwaukee, WI: U.S. Department of Agriculture, Forest Service, Eastern Region.

Climate Change Wildlife Action Plan Work Group. 2009. **Voluntary guidance for states to incorporate climate change into state wildlife action plans & other management plans.** Association of Fish & Wildlife Agencies. Available at http://files.dnr.state.mn.us/eco/swap/climate_change_guidance.pdf. (Accessed 8 September 2011).

Coakley, S.M.; Scherm, H.; Chakraborty, S. 1999. **Climate change and plant disease management.** Annual Review of Phytopathology. 37: 399-426.

Coates, D.J.; Dixon, K.W. 2007. **Current perspectives in plant conservation biology.** Australian Journal of Botany. 55(3): 187-193.

Covington, W.W. 1981. **Changes in forest floor organic matter and nutrient content following clear cutting in northern hardwoods.** Ecology. 62(1): 41-48.

Dale, V.H.; Joyce, L.A.; McNulty, S.; Neilson, R.P.; Ayres, M.P.; Flannigan, M.D.; Hanson, P.J.; Irland, L.C.; Lugo, A.E.; Peterson, C.J.; Simberloff, D.; Swanson, F.J.; Stocks, B.J.; Wotton, B.M. 2001. **Climate change and forest disturbances.** Bioscience. 51(9): 723-734.

Davis, M.B.; Shaw, R.G. 2001. **Range shifts and adaptive responses to Quaternary climate change.** Science. 292(5517): 673-679.

Doppelt, B.; Hamilton, R.; Williams, C.D.; Koopman, M.E.; Vynne, S.J. 2009. **Preparing for climate change in the Upper Willamette River Basin of Western Oregon: co-beneficial planning for communities and ecosystems.** Available at http://www.geosinstitute.org/ images/stories/pdfs/Publications/ClimateWise/ UpperWillametteBasinReport3-24-09FINAL. pdf. (Accessed 8 September 2011).

Dukes, J.S.; Pontius, J.; Orwig, D.; Garnas, J.R.; Rodgers, V.L.; Brazee, N.; Cooke, B.; Theoharides, K.A.; Stange, E.E.; Harrington, R.; Ehrenfeld, J.; Gurevitch, J.; Lerdau, M.; Stinson, K.; Wick, R.; Ayres, M. 2009. **Responses of insect pests, pathogens, and invasive plant species to climate change in the forests of northeastern North America: What can we predict?** Canadian Journal of Forest Research. 39(2): 231-248.

Duvall, M.D.; Grigal, D.F. 1999. **Effects of timber harvesting on coarse woody debris in red pine forests across the Great Lakes states, USA.** Canadian Journal of Forest Research. 29(12): 1926-1934.

ECOMAP. 1993. **National hierarchical framework of ecological units.** Washington, DC: U.S. Department of Agriculture, Forest Service. 20 p.

Fiedler, P.L.; Laven, R.D. 1996. **Selecting reintroduction sites.** In: Falk, D.A.; Millar, C.I.; Olwell, M., eds. Restoring diversity: strategies for reintroduction of endangered plants. Washington, DC: Island Press: 157-169.

Fischlin, A.; Ayres, M.; Karnosky, D.; Kellomäki, S.; Louman, B.; Ong, C.; Plattner, G.K.; Santoso, H.; Thompson, I.; Booth, T.H.; Marcar, N.; Scholes, B.; Swanston, C.; Zamolodchikov, D. 2009. **Future environmental impacts and vulnerabilities.** Helsinki, Finland: International Union of Forest Research Organizations: 53-100.

Frelich, L.E. 2002. **Forest dynamics and disturbance regimes: studies from temperate evergreen-deciduous forests.** New York, NY: Cambridge University Press. 280 p.

Frelich, L.E.; Reich, P.B. 2010. **Will environmental changes reinforce the impact of global warming on the prairie-forest border of central North America?** Frontiers in Ecology and the Environment. 8(7): 371-378.

Galatowitsch, S.; Frelich, L.; Phillips-Mao, L. 2009. **Regional climate change adaptation strategies for biodiversity conservation in a midcontinental region of North America.** Biological Conservation. 142(10): 2012-2022.

Gitay, H.; Suarez, A.; Watson, R.T.; Dokken, D.J. 2002. **Climate change and biodiversity.** Geneva, Switzerland: Intergovernmental Panel on Climate Change. 85 p.

Glick, P.; Stein, B.A.; Edelson, N.A., eds. 2011. **Scanning the conservation horizon: a guide to climate change vulnerability assessment.** Washington, DC: National Wildlife Federation. 176 p. Available at http://www.californiawildlifefoundation.org/ articles/ScanningtheConservationHorizon.pdf. (Accessed 24 August 2011).

Gregg, L.; Heeringa, B.; Ecklund, D. 2004. **Conservation assessment for spruce grouse (*Falcipennis canadensis*).** Milwaukee, WI: U.S. Department of Agriculture, Forest Service, Eastern Region. 33 p.

Groves, C.; Anderson, M.; Enquist, C.; Girvetz, E.; Sandwith, T.; Schwarz, L.; Shaw, R. 2010. **Climate change and conservation: a primer for assessing impacts and advancing ecosystem-based adaptation in The Nature Conservancy.** N.p.: The Nature Conservancy. 55 p.

Gunn, J.S.; Hagan, J.M.; Whitman, A.A. 2009. **Forestry adaptation and mitigation in a changing climate: a forest resource manager's guide for the northeastern United States.** Brunswick, ME: Manomet Center for Conservation Sciences. 16 p.

Halofsky, J.E.; Peterson, D.L.; O'Halloran, K.; Hawkins Hoffman, C., eds. 2011. **Adapting to climate change at Olympic National Forest and Olympic National Park.** Gen. Tech. Rep. PNW-GTR-844. Portland, OR: U.S. Department of Agriculture, Forest Service, Pacific Northwest Research Station. 130 p.

Halpin, P.N. 1997. **Global climate change and natural-area protection: management responses and research directions.** Ecological Applications. 7(3): 828-843.

Hannah, L. 2008. **Protected areas and climate change.** Annals of the New York Academy of Sciences. 1134 (The Year in Ecology and Conservation Biology 2008): 201-212.

Harris, J.A.; Hobbs, R.J.; Higgs, E.; Aronson, J. 2006. **Ecological restoration and global climate change.** Restoration Ecology. 14(2):170-176.

Heinz Center. 2008. **Strategies for managing the effects of climate change on wildlife and ecosystems.** 43 p. Available at http://www.heinzctr.org/publications/PDF/Strategies_for_managing_effects_of_climate_change_on_wildlife.pdf. (Accessed 25 December 2011).

Heller, N.E.; Zavaleta, E.S. 2009. **Biodiversity management in the face of climate change: a review of 22 years of recommendations.** Biological Conservation. 142(1): 14-32.

Hellmann, J.J.; Byers, J.E.; Bierwagen, B.G.; Dukes, J.S. 2008. **Five potential consequences of climate change for invasive species.** Conservation Biology. 22(3): 534-543.

Innes, J.; Joyce, L.A.; Kellomäki, S.; Louman, B.; Ogden, A.; Parrotta, J.; Thompson, I. 2009. **Management for adaptation.** In: Seppälä, R.; Buck, A.; Katila, P., eds. Adaptation of forests and people to climate change. A Global Assessment Report. IUFRO World Series 22. Helsinki, Finland: International Union of Forest Research Organizations: 135-186.

Intergovernmental Panel on Climate Change [IPCC]. 1995. **Guidance notes for lead authors of the IPCC fourth assessment report on addressing uncertainties.** Geneva, Switzerland: Intergovernmental Panel on Climate Change. Available at http://www.ipcc.ch/pdf/supporting-material/uncertainty-guidance-note.pdf. (Accessed 28 December 2011).

Intergovernmental Panel on Climate Change. 2007. **Climate Change 2007: The physical science basis: summary for policy makers.** Cambridge, UK and New York, NY: Cambridge University Press, USA. 142 p. Available at http://www.ipcc.ch/publications_and_data/ar4/wg1/en/contents.html. (Accessed 28 December 2011).

Iverson, L.R.; Schwartz, M.W.; Prasad, A.M. 2004. **How fast and far might tree species migrate in the eastern United States due to climate change?** Global Ecology and Biogeography. 13(3): 209-219.

Janowiak, M.K.; Swanston, C.W.; Nagel, L.M.; Webster, C.R.; Palik, B.J.; Twery, M.J.; Bradford, J.B.; Parker, L.R.; Hille, A.T.; Johnson, S.M. 2011. **Silvicultural decisionmaking in an uncertain climate future: a workshop-based exploration of considerations, strategies, and approaches.** Gen. Tech. Rep. NRS-81. Newtown Square, PA: U.S. Department of Agriculture, Forest Service, Northern Research Station. 14 p.

Johnson, J.E. 1995. **The Lake States Region.** In: Barrett, J.W., ed. Regional silviculture of the United States. New York, NY: John Wiley and Sons: 81-127.

Joyce, L.; Blate, G.; McNulty, S.; Millar, C.; Moser, S.; Neilson, R.; Peterson, D. 2009. **Managing for multiple resources under climate change: national forests.** Environmental Management. 44(6): 1022-1032.

Joyce, L.A.; Blate, G.M.; Littell, J.S.; McNulty, S.G.; Millar, C.I.; Moser, S.C.; Neilson, R.P.; O'Halloran, K.; Peterson, D.L. 2008. **National Forests.** In: Julius, S.H.; West, J.M. (eds.); Baron, J.S.; Joyce, L.A.; Kareiva, P.; Keller, B.D.; Palmer, M.A.; Peterson, C.H.; Scott, J.M. (authors). Preliminary review of adaptation options for climate-sensitive ecosystems and resources. A report by the U.S. Climate Change Science Program and the Subcommittee on Global Change Research. Washington, DC: U.S. Environmental Protection Agency: 3-1 to 3-127.

Kling, G.W.; Hayhoe, K.; Johnson, L.B.; Magnuson, J.J.; Polasky, S.; Robinson, S.K.; Shuter, B.J.; Wander, M.M.; Wuebbles, D.J.; Zak, D.R.; Lindroth, R.L.; Moser, S.C.; Wilson, M.L. 2003. **Confronting climate change in the Great Lakes region: impacts on our communities and ecosystems.** Cambridge, MA: Union of Concerned Scientists; Ecological Society of America. Available at http://ucsusa.org/assets/documents/global_warming/greatlakes_final.pdf. (Accessed 28 December 2011).

Kucharik, C.; Serbin, S.; Vavrus, S.; Hopkins, E.; Motew, M. 2010. **Patterns of climate change across Wisconsin from 1950 to 2006.** Physical Geography. 31(1): 1-28.

Lawler, J.J. 2009. **Climate change adaptation strategies for resource management and conservation planning.** Year in Ecology and Conservation Biology 2009. 1162: 79-98.

Lawler, J.J.; Tear, T.H.; Pyke, C.; Shaw, M.R.; Gonzalez, P.; Kareiva, P.; Hansen, L.; Hannah, L.; Klausmeyer, K.; Aldous, A.; Bienz, C.; Pearsall, S. 2010. **Resource management in a changing and uncertain climate.** Frontiers in Ecology and the Environment. 8(1): 35-43.

Levina, E.; Tirpak, D. 2006. **Adaptation to climate change: key terms.** Paris, France: Organisation for Economic Co-operation and Development and International Energy Agency. 24 p. Available at http://www.oecd.org/dataoecd/36/53/36736773.pdf. (Accessed 28 December 2011).

Lindenmayer, D.B.; Franklin, J.F.; Fischer, J. 2006. **General management principles and a checklist of strategies to guide forest biodiversity conservation.** Biological Conservation. 131(3): 433-445.

MacDonald, G.M.; Cwynar, L.C.; Whitlock, C. 1998. **The late Quaternary dynamics of pines in northern North America.** In: Richardson, D.M., ed. Ecology and biogeography of *Pinus*. Cambridge, UK: Cambridge University Press: 122-136.

Manomet Center for Conservation Sciences. 2010. **Climate change and Massachusetts fish and wildlife: volume 1.** Brunswick, ME: Manomet Center for Conservation Sciences. 19 p.

McLachlan, J.S.; Hellmann, J.J.; Schwartz, M.W. 2007. **A framework for debate of assisted migration in an era of climate change.** Conservation Biology. 21(2): 297-302.

Millar, C.I. 1991. **Conservation of germplasm in forest trees.** In: Ahuja, M.R.; Libby, W.J., eds. Clonal forestry II. Conservation and application. New York, NY: Springer-Verlag: 42-65.

Millar, C.I.; Brubaker, L.B. 2006. **Climate change and paleoecology: new contexts for restoration ecology.** In: Palmer, M; Falk, D; Zedler, J., eds. Restoration science. Washington, DC: Island Press: 315-340.

Millar, C.I.; Stephenson, N.L.; Stephens, S.L. 2007. **Climate change and forests of the future: managing in the face of uncertainty.** Ecological Applications. 17(8): 2145-2151.

Millar, C.I.; Stephenson, N.L.; Stephens, S.L. 2008. **Re-framing forest and resource management strategies for a climate change context.** Mountain views: the newsletter of the Consortium for Integrated Climate Research in Western Mountains. 2(1): 5-10. Available at http://www.fs.fed.us/psw/cirmount/publications/pdf/Mtn_Views_feb_08.pdf. (Accessed 28 December 2011).

Mladenoff, D.J.; Stearns, F. 1993. **Eastern hemlock regeneration and deer browsing in the northern Great Lakes region: a re-examination and model simulation.** Conservation Biology. 7(4): 889-900.

Morelli, T.L.; Yeh, S.; Smith, N.; Millar, C.I. In press. **Climate project screening tool: an aid for climate change adaptation.** Gen. Tech. Rep. Portland, OR: U.S. Department of Agriculture, Forest Service, Pacific Northwest Research Station.

National Research Council. 2010. **Adapting to the impacts of climate change.** Washington, DC: The National Academies Press. 292 p.

Northern Institute of Applied Climate Science. 2011. **The Climate Change Response Framework Project in northern Wisconsin.** Available at http://www.nrs.fs.fed.us/niacs/climate/northwoods/ (Accessed 27 April 2012).

Noss, R.F. 2001. **Beyond Kyoto: forest management in a time of rapid climate change.** Conservation Biology. 15(3): 578-590.

Ogden, A.; Innes, J. 2008. **Climate change adaptation and regional forest planning in southern Yukon, Canada.** Mitigation and Adaptation Strategies for Global Change. 13(8): 833-861.

Parry, M.L.; Canziani, O.F.; Palutikof, J.P. 2007. **Climate change 2007: impacts, adaptation and vulnerability.** Technical Summary. Cambridge, UK: Intergovernmental Panel on Climate Change: 23-78. Available at http://www.ipcc.ch/ipccreports/ar4-wg2.htm. (Accessed 8 September 2011).

Peterson, D.L.; Millar, C.I.; Joyce, L.A.; Furniss, M.J.; Halofsky, J.E.; Neilson, R.P.; Morelli, T.L. In press. **Responding to climate change on National Forests: a guidebook for developing adaptation options.** Gen. Tech. Rep. Portland, OR: U.S. Department of Agriculture, Forest Service, Pacific Northwest Research Station.

Peterson, C.J. 2000. **Catastrophic wind damage to North American forests and the potential impact of climate change.** The Science of the Total Environment. 262: 287-311.

Post, E. 2003. **Climate-vegetation dynamics in the fast lane.** Trends in Ecology & Evolution. 18(11): 551-553.

Prasad, A.M.; Iverson, L.R.; Matthews, S.N.; Peters, M. 2007. **A climate change atlas for 134 forest tree species of the eastern United States [database].** Delaware, OH: U.S. Department of Agriculture, Forest Service, Northern Research Station. Available at http://www.nrs.fs.fed.us/atlas/tree. (Accessed 26 December 2011).

Rhemtulla, J.M.; Mladenoff, D.J.; Clayton, M.K. 2009. **Legacies of historical land use on regional forest composition and structure in Wisconsin, USA (mid-1800s-1930s-2000s).** Ecological Applications. 19(4): 1061-1078.

Rooney, T.P.; Waller, D.M. 2003. **Direct and indirect effects of white-tailed deer in forest ecosystems.** Forest Ecology and Management. 181(1-2): 165-176.

Schiermeier, Q. 2010. **The real holes in climate science.** Nature. 463(7279): 284-287.

Secretariat of the Convention on Biological Diversity. 2009. **Connecting biodiversity and climate change mitigation and adaptation: report of the second ad hoc technical expert group on biodiversity and climate change.** Technical Series No. 41. Montreal, QUE: Convention on Biological Diversity. 126 p. Available at http://www.cbd.int/doc/publications/cbd-ts-41-en.pdf. (Accessed 26 December 2011).

Society of American Foresters. 2011. **The dictionary of forestry.** Bethesda, MD: Society of American Foresters. Available at: http://dictionaryofforestry.org/. (Accessed 8 August 2011).

Spittlehouse, D.L. 2005. **Integrating climate change adaptation into forest management.** Forestry Chronicle. 81(5): 691-695.

Spittlehouse, D.L.; Stewart, R.B. 2003. **Adaptation to climate change in forest management.** BC Journal of Ecosystems and Management. 4(1): 1-11.

Stankey, G.H.; Clark, R.N.; Bormann, B.T. 2005. **Adaptive management of natural resources: theory, concepts, and management institutions.** Gen. Tech. Rep. PNW-GTR-654. Portland, OR: U.S. Department of Agriculture, Forest Service, Pacific Northwest Research Station. 73 p.

Swanston, C.; Janowiak, M.; Iverson, L.; Parker, L.; Mladenoff, D.J.; Brandt, L.; Butler, P.; St. Pierre, M.; Prasad, A.; Matthews, S.; Peters, M.; Higgins, D.; Dorland, A. 2011. **Ecosystem vulnerability assessment and synthesis: a report from the Climate Change Response Framework Project in northern Wisconsin.** Gen. Tech. Rep. NRS-82. Newtown Square, PA: U.S. Department of Agriculture, Forest Service, Northern Research Station. 142 p.

The Nature Conservancy. 2009. **Conservation action planning guidelines for developing strategies in the face of climate change.** Available at http://conserveonline.org/workspaces/climateadaptation/documents/climate-change-project-level-guidance. (Accessed 8 September 2011).

Tirmenstein, D.A. 1991. *Acer saccharum.* In: Fire Effects Information System, [database]. Fort Collins, CO: U.S. Department of Agriculture, Forest Service, Rocky Mountain Research Station, Fire Sciences Laboratory. Available at http://www.fs.fed.us/database/feis/plants/tree/acesac/introductory.html. (Accessed 8 September 2011).

Union of Concerned Scientists. 2009. **Confronting climate change in the U.S. Midwest.** Cambridge, MA: Union of Concerned Scientists. Series available at http://www.ucusa.org/global_warming/science_and_impacts/impacts/climate-change-midwest.html. (Accessed 8 January 2012).

U.S. Forest Service. 2008. **Forest Service strategic framework for responding to climate change.** Available at http://www.fs.fed.us/climatechange/documents/strategic-framework-climate-change-1-0.pdf (Accessed 28 December 2011).

U.S. Forest Service. 2010. **National roadmap for responding to climate change.** Available at http://www.fs.fed.us/climatechange/advisor/roadmap.html. (Accessed 28 December 2011).

U.S. Global Change Research Program. 2009. **Global climate change impacts in the United States.** Washington, DC: U.S. Global Change Research Program. 188 p. Available at http://www.globalchange.gov/publications/reports/scientific-assessments/us-impacts. (Accessed 8 January 2012).

Waller, D. 2007. **White-tailed deer impacts in North America and the challenge of managing a hyperabundant herbivore.** In: Gaston, A.J.; Golumbia, T.E.; Martin, J.L.; Sharpe, S.T., eds. Lessons from the Islands—introduced species and what they tell us about how ecosystems work. Proceedings from the Research Group on Introduced Species 2002 Symposium, Queen Charlotte City, Queen Charlotte Islands, BC. Special Publication. Ottawa, ONT: Canadian Wildlife Service, Environment Canada: 135-147.

Waller, D.M.; Alverson, W.S. 1997. **The white-tailed deer: a keystone herbivore.** Wildlife Society Bulletin. 25(2): 217-226.

Wisconsin Department of Natural Resources. 1995. **Wisconsin's biodiversity as a management issue- a report to Department of Natural Resources managers.** Available at http://dnr.wi.gov/org/es/science/publications/rs915_95.htm. (Accessed 8 September 2011).

Wisconsin Department of Natural Resources. 1998. **Land cover data (WISCLAND).** Available at http://dnr.wi.gov/maps/gis/datalandcover.html. (Accessed 8 September 2011).

Wisconsin Department of Natural Resources . 2009. **Wisconsin's forestry best management practices invasive species.** Available at http://council.wisconsinforestry.org/invasives/pdf/FinalForestryBMPManual_03-26-09.pdf. (Accessed 8 September 2011).

Wisconsin Department of Natural Resources. 2010. **Silviculture handbook 2431.5.** Available at http://dnr.wi.gov/forestry/publications/handbooks/24315/24315.pdf. (Accessed 18 August 2011).

Wiens, J.A.; Stralberg, D.; Jongsomjit, D.; Howell, C.A.; Snyder, M.A. 2009. **Niches, models, and climate change: assessing the assumptions and uncertainties.** Proceedings of the National Academy of Sciences. 106 (Supplement 2): 19729-19736.

Wisconsin Initiative on Climate Change Impacts. 2011a. **Climate Working Group interactive mapping tool.** Madison, WI: Board of Regents of the University of Wisconsin System. Available at http://www.wicci.wisc.edu/climate-change.php. (Accessed 17 April 2012).

Wisconsin Initiative on Climate Change Impacts. 2011b. **Wisconsin's changing climate: impacts and adaptation.** Madison, WI: Nelson Institute for Environmental Studies, University of Wisconsin-Madison and the Wisconsin Department of Natural Resources. 217 p.

Woodall, C.W.; Oswalt, C.M.; Westfall, J.A.; Perry, C.H.; Nelson, M.D.; Finley, A.O. 2009. **An indicator of tree migration in forests of the eastern United States.** Forest Ecology and Management. 257(5): 1434-1444.

Worland, M.; Martin, K.J.; Gregg, L. 2009. **Spruce grouse distribution and habitat relationships in Wisconsin.** The Passenger Pigeon. 71: 5-18.

APPENDIX 1. MONITORING NETWORKS IN THE GREAT LAKES REGION

Network	Research Topics	Established
Ameriflux	Air temperature, carbon dioxide and water vapor concentrations, photosynthetically active radiation, leaf wetness, dew point temperature, relative humidity, soil moisture, global solar radiation, and wind direction and speed. http://webmap.ornl.gov/daac/viewer.htm?SOURCE=FLUXNET	1996
Breeding Bird Survey	Bird populations. http://www.mbr-pwrc.usgs.gov/bbs/bbs.html	1966
CASTNET	Nitrogen and sulfur deposition, air pollutants, visibility/fine particulates, and toxins. http://www.epa.gov/castnet/sites/prk134.html	1988
Chequamegon Ecosystem-Atmosphere Study (ChEAS)	Biosphere/atmosphere interactions within a northern mixed forest in northern Wisconsin. A primary goal is to understand the processes controlling forest-atmosphere exchange of carbon dioxide and the response of these processes to climate change. Another primary goal is to bridge the gap between canopy-scale flux measurements and the global CO_2 flask-sampling network. http://cheas.psu.edu/	Varies by item monitored
Ecological Data Wiki	This website is a hub for datasets across the United States. These datasets are searchable by taxon, biome, data type, ecological level (species, population, community, and ecosystem), and spatial extent. http://ecologicaldata.org/	2011
Field Sampled Vegetation (FSVeg)	U.S. Forest Service inventory system that contains data about trees, fuels, down woody material, surface cover, and understory vegetation. FSVeg supports the business of common stand exams, fuels data collection, permanent grid inventories, and other vegetation inventory collection processes. http://www.fs.fed.us/emc/nris/products/fsveg/index.shtml	Varies by item monitored
Forest Activities Tracking System (FACTS)	An activity-tracking system for all levels of the Forest Service. It supports timber sales in conjunction with Timber Information Manager Contracts and Permits; tracks and monitors decisions under the National Environmental Policy Act; tracks Knutson-Vandenberg trust fund plans at the timber sale level and reporting at the National level; and generates National, Regional, Forest, and/or District reports.	Varies by item monitored

(continued on next page)

Network	Research Topics	Established
Forest Birds of the Western Great Lakes	Long-term monitoring program with more than 1,600 off-road sampling points designed to track regional population trends and investigate forest birds' response to regional land use patterns. http://www.nrri.umn.edu/mnbirds/fbd_forms0.asp	1992
Forest Inventory and Analysis (FIA)	The U.S. Forest Service FIA database contains extensive data on forest area attributes, such as forest type, stand size, stand age, and forest disturbance. Data are also collected on individual trees, which are tracked over time. Tree attributes include species identification, diameter at breast height, tree length, and tree damage. More in-depth measurements are collected on a subsample of the plots: tree crown condition, lichen communities, ozone injury, down woody materials, soil condition, and vegetation diversity and structure. http://fia.fs.fed.us/tools-data/other/default.asp	Periodically, 1930-1998; annually, 1999
Forest Service Natural Resource Information System (NRIS)	Biological, physical, and human features that make up National Forest and Grassland landscapes. http://www.fs.fed.us/emc/nris/products/index.shtml	Varies by item monitored
Great Lakes Network Inventory and Monitoring Program	The National Park Service has organized 32 'networks' of parks across the nation to monitor a wide range of items, including: water and air quality; terrestrial plants; weather; succession; land cover/use; terrestrial pests and pathogens; soils; aquatic/wetland plant communities; plant and animal exotics; amphibians; stream dynamics; birds; diatoms; fish; trophic bioaccumulation; water level and flow; and species health, growth and reproductive success. http://science.nature.nps.gov/im/units/GLKN/index.cfm	Varies by item monitored
Long Term Ecological Research - North Temperate Lakes (LTER – NTL)	Physical, chemical and biological limnology; hydrology and geochemistry; paleolimnology; climate forcing; producer and consumer ecology; ecology of invasions; ecosystem variability; landscape ecology; lake, landscape and human interactions. http://www.lternet.edu/sites/ntl/	1981
National Atmospheric Deposition Program (NADP)	The Mercury Deposition Network (MDN) provides a long-term record of total mercury (Hg) concentration and deposition in precipitation; the Atmospheric Integrated Research Monitoring Network (AIRMoN) reports daily measurements of the acids, nutrients, and base cations in precipitation for studying and modeling atmospheric processes; the Atmospheric Mercury Network (AMNet) reports atmospheric Hg concentrations for determination of mercury dry deposition; the National Trends Network (NTN) provides a long-term record of the acids, nutrients, and base cations in U.S. precipitation. http://nadp.sws.uiuc.edu/NADP/	1978
National Ecological Observatory Network (NEON)	NEON has partitioned the United States into 20 eco-climatic domains, each of which represents different regions of vegetation, landforms, climate, and ecosystem performance. In those domains, NEON will collect site-based data about climate and atmosphere, soils and streams and ponds, and a variety of organisms. Additionally, NEON will provide a wealth of regional and national-scale data from airborne observations and geographical data collected by Federal agencies and processed by NEON to be accessible and useful to the ecological research community. NEON will also manage a long-term multi-site stream experiment and provide a platform for future observations and experiments proposed by the scientific community. http://www.neoninc.org/domains/greatlakes	Under development

(continued on next page)

Network	Research Topics	Established
National Streamflow Information Program (NSIP)	Data transmitted from selected surface-water, groundwater, and water-quality sites. Real-time data typically are recorded at 15- to 60-minute intervals, stored at the gaging station, and then transmitted to U.S. Geological Survey offices every few hours. http://wdr.water.usgs.gov/nwisgmap/?state=wi	1889
National Woodland Ownership Survey	Demographics of private landowners; results from survey questions on forest acquisition, forest holdings, management activities, and timber harvests. http://fiatools.fs.fed.us/NWOS/tablemaker.jsp	2002
Natural Resources Inventory (NRI)	A statistical survey of land use and natural resource conditions and trends on U.S. non-Federal lands: Air visibility/fine particulates; water sediment load; soil texture, chemistry, toxicity, mineralogy, climate, structure, strength, erodability; vegetation growth rate/above ground biomass, species/cover/range; management activities. http://www.nrcs.usda.gov/technical/NRI/	1956
Nicolet National Forest Bird Survey	Records of all birds seen and heard from a single point (survey site) during a 10-minute period. http://www.uwgb.edu/birds/nnf/	1987
USA National Phenology Network	Recurring plant and animal life cycle stages, or phenophases, such as leafing and flowering of plants, maturation of agricultural crops, emergence of insects, and migration of birds. http://www.usanpn.org/home	Varies by item monitored

APPENDIX 2. SYNTHESIS OF ADAPTATION STRATEGIES AND APPROACHES

There is a wealth of literature on climate change adaptation, but most of it is quite conceptual and not designed for use at the scales most relevant to land managers. By understanding which strategies and approaches are best suited to different forest types in northern Wisconsin, we can help inform managers and allow them to develop tactics for specific site conditions, management objectives, and anticipated climate change vulnerabilities. The adaptation strategies and approaches described in Chapter 2 of this document are intended to provide a "menu" of actions to respond to climate change and make forested ecosystems in northern Wisconsin more adapted to future conditions. The methods and processes used to develop and refine this "menu" are described below.

Synthesis of Adaptation Literature

We compiled a list of climate change adaptation actions from numerous sources both in peer-reviewed and the gray literature (Table 1). Because the amount of information on adaptation is constantly increasing, we included several review papers (e.g., Heller and Zavaleta 2009, The Nature Conservancy 2009) that summarized adaptation actions from a number of other sources. The actions found in these papers were organized into strategies for adaptation as well as more specific approaches; refer to the continuum figure (Fig. 6) in Chapter 2. These strategies and approaches were nested under one or more of the broad adaptation options

of resistance, resilience, and response described by Millar et al. (2007). Through this process, a list of more than 100 adaptation actions found during the literature review was developed into a preliminary list of 10 strategies and 43 approaches.

Expert Comment Process

We sought to draw upon the experience and expertise of forest ecologists, researchers, and land managers to help us understand how these strategies and approaches relate to the forest types present in northern Wisconsin. We compiled a list of individuals with expertise in regional forest ecology and management, and 31 of the experts agreed to comment. We assigned individuals to 1 of 12 forest types (Table 1) based upon their interests, expertise, and experience. At least two experts discussed each forest type.

We elicited expert comments using a survey that walked the experts through the process of evaluating all of the adaptation strategies and approaches for a single forest type. Short descriptions of the 10 adaptation strategies and 43 adaptation approaches were provided. We recommended that before beginning the survey, experts should spend some time thinking about potential climate change impacts that are most important to the forest type they would be considering, with the *Ecosystem Vulnerability Assessment and Synthesis* (Swanston et al. 2011) as one potential resource.

Table 1.—Literature reviewed for compilation of adaptation strategies and approaches. Full citations for these resources are in the Literature Cited section of this document.

Biringer 2003. **Forest ecosystems threatened by climate change: promoting long-term forest resilience.**

Brandt et al. In Review. **Climate change science applications and needs in forest ecosystem management.**

Climate Change Wildlife Action Plan Work Group, Association of Fish and Wildlife Agencies 2009. **Voluntary guidance for states to incorporate climate change into state wildlife action plans and other management plans.**

Dale et al. 2001. **Climate change and forest disturbances.**

Galatowitsch et al. 2009. **Regional climate change adaptation strategies for biodiversity conservation in a midcontinental region of North America.**

Gitay et al. 2002. **Climate Change and Biodiversity.**

Gunn et al. 2009. **Forestry adaptation and mitigation in a changing climate: a forest resource manager's guide for the northeastern United States.**

Halpin 1997. **Global climate change and natural-area protection: management responses and research directions.**

Heinz Center 2008. **Strategies for managing the effects of climate change on wildlife and ecosystems.**

Heller and Zavaleta 2009. **Biodiversity management in the face of climate change: a review of 22 years of recommendations.**

Innes et al. 2009. **Management for adaptation.**

Joyce et al. 2009. **Managing for multiple resources under climate change: National Forests.**

Lawler 2009. **Climate change adaptation strategies for resource management and conservation planning.**

Lawler et al. 2010. **Resource management in a changing and uncertain climate.**

Lindenmayer et al. 2006. **General management principles and a checklist of strategies to guide forest biodiversity conservation.**

Manomet Center for Conservation Sciences, and Massachusetts Division of Fisheries and Wildlife 2010. **Climate change and Massachusetts fish and wildlife: volume 1.**

McLachlan et al. 2007. **A framework for debate of assisted migration in an era of climate change.**

Millar et al. 2007. **Climate change and forests of the future: managing in the face of uncertainty.**

Millar et al. 2008. **Re-framing forest and resource management strategies for a climate change context.**

Spittlehouse 2005. **Integrating climate change adaptation into forest management.**

Spittlehouse and Stewart 2003. **Adaptation to climate change in forest management.**

The Nature Conservancy 2009. **Conservation action planning guidelines for developing strategies in the face of climate change.**

U.S. Forest Service 2008. **Forest Service strategic framework for responding to climate change.**

U.S. Forest Service 2010. **National roadmap for responding to climate change.**

For each adaptation strategy, experts were first asked a set of questions about the applicability of each adaptation approach under that strategy for the forest type in question. Experts were then asked to consider how effectively the suite of approaches helped support the broader strategy. The questions were open-ended, and the survey format helped us to effectively collate answers and perspectives from multiple reviewers. Experts were asked to evaluate every strategy and every approach. If a strategy or approach was not useful in the forest type considered, we wanted to know why it does not work. We also asked the experts to keep the following in mind while answering the questions: (1) Forest types are broad groupings, and management decisions will be highly dependent upon location, site conditions, management objectives, or other factors; and (2) Despite social, economic, and other constraints, we want to focus at this time on the capability of ecosystems; options that are not socially or economically viable should not be discounted at this time.

For each approach, we asked the following questions:

1) How well would this approach work for the forest type in question?
2) What is special about this forest type that would affect the use of this approach in northern Wisconsin?
3) Are there situations where you would recommend or discourage use of this approach in this forest type?

For each strategy, we asked the following questions:

1) After reviewing the approaches under this strategy, are there additional approaches that you would recommend?
2) After reviewing all of the approaches under this strategy, how well does this strategy work overall?

Experts were given 2 weeks to respond. After the deadline, survey responses were collated by forest type, then by strategy, and then by approach, and comments from multiple reviewers were listed together. Responses for all approaches and strategies were evaluated and discussed at a meeting of the authors, and a revised list of 10 strategies and 41 approaches was created.

Feedback from Illustration Teams

In fall 2010, we worked with two teams of land managers from the Chequamegon-Nicolet National Forest to develop the Illustrations presented in this document (Chapter 4). Each team included four or five land managers of different specialties (e.g., silviculture, wildlife, hydrology). We worked with each team to define a unique management issue of interest and held a series of meetings where the managers used the revised list of Adaptation Strategies and Approaches to complete a preliminary version of the Adaptation Workbook. We closely observed the teams as they completed the Adaptation Workbook, and recorded detailed notes of the teams' discussions as they selected adaptation approaches and devised adaptation tactics. Based on comments from the teams, the approaches were further refined and condensed into 39 approaches under 10 strategies. Additionally, comments from the teams were used to refine the forest type-specific considerations listed under each approach.

APPENDIX 3. LIST OF COMMON AND SCIENTIFIC NAMES

FLORA

Common Name	Scientific Name
balsam fir	*Abies balsamea*
red maple	*Acer rubrum*
sugar maple	*Acer saccharum*
yellow birch	*Betula alleghaniensis*
paper birch	*Betula papyrifera*
white ash	*Fraxinus americana*
black ash	*Fraxinus nigra*
tamarack	*Larix laricina*
eastern hophornbeam	*Ostrya virginiana*
Norway spruce	*Picea abies*
white spruce	*Picea glauca*
black spruce	*Picea mariana*
jack pine	*Pinus banksiana*
red pine	*Pinus resinosa*
eastern white pine	*Pinus strobus*
Scotch pine	*Pinus sylvestris*
balsam poplar	*Populus balsamifera*

FLORA (continued)

Common Name	Scientific Name
bigtooth aspen	*Populus grandidentata*
quaking aspen	*Populus tremuloides*
black cherry	*Prunus serotina*
white oak	*Quercus alba*
northern pin oak	*Quercus ellipsoidalis*
northern red oak	*Quercus rubra*
northern white-cedar	*Thuja occidentalis*
American basswood	*Tilia americana*
eastern hemlock	*Tsuga canadensis*
American elm	*Ulmus americana*

FAUNA

Common Name	Scientific Name
emerald ash borer	*Agrilus planipennis*
spruce grouse	*Falcipennis canadensis*
white-tailed deer	*Odocoileus virginianus*

APPENDIX 4. PAPER BIRCH FOREST ILLUSTRATION

A more complete set of information is provided below for the Paper Birch Forest Illustration (see Chapter 4 for a description of the illustration). Because the Adaptation Workbook was modified based upon the illustration teams' comments, the information from the teams has been modified to fit the current Adaptation Workbook structure and to better provide managers with an example that can be consulted while completing the Adaptation Workbook. The following tables contain our interpretation of the ideas, issues, and responses that were developed by the illustration team.

Paper birch trees.

Worksheet #1/Step #1: DEFINE area of interest, management goals and objectives, and time frames.

Area of Interest	Location	Forest Type(s)	Management Goals	Management Objectives	Time Frames
Early-successional paper birch forest within the defined project area (in Management Area 1B: Early Successional Aspen, Mixed Aspen-Conifer, and Conifer) *More information on this Management Area is available on pages 3-4 and 3-5 of the 2004 Land and Resource Management Plan (Chequamegon-Nicolet National Forest 2004).*	Paper birch stands within the project area	Paper birch	1) Retain paper birch forest on the landscape. 2) Increase species and structural diversity.	1) Regenerate the existing mature paper birch to retain it on the landscape when desirable. 2) Regenerate or underplant white pine among the natural paper birch when: (1) opportunities exist to improve stand diversity, (2) paper birch regeneration isn't possible, or (3) site scarification is not possible or desired.	Implementation is expected in the immediate future (2 years or less). Many management goals will be realized in the long term as paper birch is regenerated. The end of the next rotation is in ~60 years.

APPENDIX 4. PAPER BIRCH FOREST ILLUSTRATION; STEP #2

Worksheet #2/Step #2: ASSESS climate change impacts and vulnerabilities for the area of interest.

Broad-scale Impacts and Vulnerabilities	Climate Change Impacts and Vulnerabilities for the Area of Interest
	How might broad-scale impacts and vulnerabilities be affected by conditions in <u>your area of interest?</u> • Landscape pattern • Site location, such as topographic position or proximity to water features • Soil characteristics • Management history or current management plans • Species or structural composition • Presence of or susceptibility to pests, disease, or nonnative species that may become more problematic under future climate conditions • Other.....
Warmer temperatures	
Longer growing seasons	
Altered precipitation regimes	
Drier soils during summer	Drier soils during summer and increased potential for drought. Warmer and drier conditions may be a substantial challenge because paper birch is on the edge of its range.
Increased threats from insects, diseases, and invasive plants	In addition to increased threat of non-native invasives, native species like raspberry can impede paper birch regeneration.
Altered disturbance regimes may lead to changes in successional trajectories	Altered disturbance regimes may cause successional changes or large and severe loss of forest cover. Medium-scale disturbances (e.g., wind) are likely to accelerate succession to another forest type.
Many common tree species are projected to have reduced habitat suitability	Many tree species are projected to have reduced habitat suitability, including paper birch, aspen, balsam fir, and other common boreal species that are near the southern extent of their range. Suitable habitat of paper birch is expected to decline as a result of climate change and may accelerate the current decline of that species in northern Wisconsin. Pine species are somewhat less vulnerable. Oak species may be favored.
Decline of associated rare species	Decline of associated rare species
Decline of associated wildlife species	Decline of associated wildlife species
Increased fire and wind disturbance	Regeneration of paper birch is difficult because fire cannot be used in most stands (and other site preparation methods can also be constrained by such factors as visual concerns and topography). Hotter, drier conditions may increase probability of natural fire, but it is unlikely to occur at the desired time and place.
Increased disturbances may accelerate current decline	Increases in wind and other medium-scale disturbances may interrupt paper birch regeneration stages and contribute to the current decline.
Wind or other medium-scale disturbances may not adequately allow for reestablishment	Habitat suitability is projected to be reduced for many tree species, including paper birch, aspen, balsam fir, and other common associates. Pine species are somewhat less vulnerable. Oak species may be favored.
	Failure to re-sprout after deer browse impedes paper birch regeneration. Deer numbers may increase with warmer temperatures and milder winters. Regeneration may be more difficult in this area because the mesic site conditions are favorable to other species and may lead to increased competition.
	Because paper birch in the area is over-mature, it is more susceptible to all stressors (e.g., drought, insects) and their interactions.

101

Worksheet #3/Step #3: EVALUATE management objectives given projected impacts and vulnerabilities.

Management Objective (from Worksheet #1, column 5)	Challenges to Meeting Management Objective with Climate Change	Opportunities for Meeting Management Objective with Climate Change	Feasibility of Meeting Objective under Current Management	Other Considerations
Regenerate the existing mature paper birch to retain it on the landscape when desirable.	• Warmer temperatures and drier conditions will make it increasingly difficult to regenerate paper birch. • There is potential for more rapid decline of the species due to the northward shift in range and the expected increase in stressors. • Prescribed burn windows may become less available if fire danger is elevated, making fire site preparation more difficult to achieve.	• Increased wildfire occurrence may benefit paper birch regeneration if fire occurs under the right conditions. • If sites become drier, species that compete with paper birch regeneration may be reduced on some sites. • Beyond this area of interest, some hardwood stands may become more conducive to paper birch management in the future as site conditions change.	Short-term: Moderate Long-term: Low	• Native American tribes are interested in maintaining paper birch bark sources for baskets, canoes, and other uses. • Social resistance to prescribed burning for site preparation may increase, especially if wildfire occurrence increases. • Additional resources and support may be needed to perpetuate the species in the future; it is unknown whether these will be available at that time.
Regenerate or underplant white pine among the natural paper birch when: (1) opportunities exist to improve stand diversity, (2) paper birch regeneration is not possible, or (3) site scarification is not possible or desired	• Regeneration of white pine may become more difficult due to deer browse, competition from raspberry and other species, dry site conditions, and insect and disease outbreaks. • Premature losses of the shelterwood overstory from wind disturbance may increase white pine susceptibility to white pine weevil and other pests. • The white pine stock that is planted now may not be adapted to future conditions.	• Regeneration of white pine may become easier if site conditions become more favorable for white pine and less favorable for competition. • It may be better to regenerate white pine now because conditions in the future (50+ years) are uncertain and may be less favorable. • Beyond this area of interest, some hardwood stands may become more conducive to paper birch management in the future as site conditions change.	Short-term: High Long-term: Moderate	• During conversion of paper birch to white pine, the Forest Plan guidelines on species composition need to be considered. • White pine seedlings need to be protected from deer browse for 5–10 years after planting. • When available, opportunities should be considered to diversify stands with species that may be favored in the future, such as oak species.

Worksheet #4/Step #4: IDENTIFY adaptation approaches and tactics for implementation.

Adaptation Approach	Tactic	Time Frames	Benefits	Drawbacks and Barriers	Practicability of Tactic	Recommend Tactic?
Maintain or improve the ability of forests to resist pests and pathogens.	Treat selected over-mature paper birch stands with a shelterwood harvest followed by prescribed burning or mechanical site preparation. Prioritize the stands to be treated using a field check of site conditions.	Immediate (2 years or less)	• Younger birches tend to be less vulnerable and more resilient to many stressors. • This tactic addresses multiple challenges identified in Step #3. • Regenerating paper birch helps meet goals and objectives set out in the Forest Plan. • The primary project purpose and the need for paper birch are addressed.	• Regeneration is not guaranteed after treatment. • It is uncertain whether this approach reduces paper birch's long-term vulnerability to climate change. • Success is often dependent upon site conditions. • Paper birch regeneration is difficult because a large amount of site preparation is needed. • Sites suitable for paper birch regeneration treatments are often limited because of accessibility, aesthetic concerns, topographic limitations, and other factors.	Short-term: Moderate Long-term: Low	Yes
	Treat selected over-mature paper birch stands that have an existing white pine seed source or advance regeneration with a shelterwood harvest and scarify for white pine. Underplant white pine to augment advance regeneration if needed. Retain the overstory.	Immediate (2 years or less)	• This tactic maintains a desired forest type (white pine) in stands where paper birch regeneration is not possible or not desired. • White pine is expected to fare better than paper birch under climate change. • Stands will have a more diverse mix of tree species, which may increase management options in the future. • Regenerating these stands and enhancing white pine helps meet goals and objectives set out in the Forest Plan. •	• Paper birch will not be a dominant species in these stands in the future, but rather a component.	High	Yes

(Worksheet #4 continued on next page.)

103

Worksheet #4/Step #4 (continued): IDENTIFY adaptation approaches and tactics for implementation.

Adaptation Approach	Tactic	Time Frames	Benefits	Drawbacks and Barriers	Practicability of Tactic	Recommend Tactic?
Maintain or improve the ability of forests to resist pests and pathogens.	Treat selected over-mature paper birch stands with a shelterwood harvest followed by Bracke scarification and underplanting of white pine. Leave the overstory.	Immediate (2 years or less)	• This tactic maintains a desired forest type (white pine) in stands where paper birch regeneration is not possible or not desired. • White pine is expected to fare better than paper birch under climate change. • Stands will have a more diverse mix of tree species, which may increase management options in the future. • Regenerating these stands and enhancing white pine help meet goals and objectives set out in the Forest Plan.	• Paper birch will not be a dominant species in these stands in the future, but rather a component.	High	Yes
	Regenerate paper birch and establish white pine on a range of site conditions.	Immediate (2 years or less)	• Given uncertainty about future site conditions, helps spread risk by maintaining desired forest types across a range of sites. • Diversifying across a range of sites increases the chances for successful short-term paper birch regeneration as well as for long-term persistence.	• There may not be a wide variety of site conditions within the area, so opportunities to implement this tactic may be limited.	Moderate	Yes

Worksheet #4/Step #4 (continued): IDENTIFY adaptation approaches and tactics for implementation.

Adaptation Approach	Tactic	Time Frames	Benefits	Drawbacks and Barriers	Practicability of Tactic	Recommend Tactic?
Maintain or improve the ability of forests to resist pests and pathogens.	Adjust rotation age lengths to achieve age class distribution goals described in the Land and Resource Management Plan.	Long-term (30 years or more)	• This tactic addresses the current issue: nearly all stands are at the end of rotation age (>60 years). • Regenerating paper birch helps meet goals and objectives set out in the Forest Plan regarding age class distributions. • This action creates diversity in age classes across the landscape, which may make stands less susceptible to some climate change impacts.	• Because stands need to be regenerated very soon to maintain paper birch as a dominant species, this diversification has to occur in the next rotation. • The desired age class balance for paper birch is off-balance (nearly all stands in this area are >60 years old); many entries will be needed to achieve the desired age diversity. • Paper birch regeneration is difficult because a large amount of site preparation is needed. • Sites suitable for paper birch regeneration treatments are often limited because of accessibility, aesthetic concerns, topographic limitations, or other factors.	Moderate	Yes
Prevent the introduction and establishment of invasive plant species and remove existing invasives.	Use the recognized nonnative invasive species (NNIS) introduction and spread tactics in the NNIS guide, contract provisions, and other applicable sources (e.g., equipment cleaning requirements, weed-free seed).	Immediate (2 years or less)	• This guidance already exists. • Likelihood of regeneration of paper birch and other desirable forest species is increased. • A higher-quality ecosystem is expected to result from these actions.	• Some approaches may require increased labor or financial resources.	High	Yes

(Worksheet #4 continued on next page.)

Worksheet #4/Step #4 (continued): IDENTIFY adaptation approaches and tactics for implementation.

Adaptation Approach	Tactic	Time Frames	Benefits	Drawbacks and Barriers	Practicability of Tactic	Recommend Tactic?
Prevent the introduction and establishment of invasive plant species and remove existing invasives.	Use stocking surveys and NNIS surveys to determine whether regeneration is being adversely affected by native and nonnative invasive plant species. If it is, consider herbicide treatments of the invasives, additional scarification, and/or individual tree releases.	Immediate, short-term, medium-term (30 years or less)	• Higher likelihood of regeneration of paper birch and other desirable forest species. • Overall higher-quality ecosystem expected to result.	• Likely to require labor-intensive approaches • Dependent on availability of financial and labor resources • Possibly other drawbacks to prevention and control of invasives (e.g., effects from use of herbicides).	Moderate	Yes
	Ensure adequate overstory is retained so as not to encourage the establishment of sun-loving invasives.	Short-term (2-10 years)	• Retaining adequate canopy may make conditions less favorable to raspberry or other species that can compete with paper birch regeneration.	• The amount of canopy cover needed to deter raspberry is not precisely known, and varies by site and post-harvest conditions. • Retaining too much canopy cover can negatively affect paper birch regeneration.	Moderate	Yes
Manage herbivory to protect or promote regeneration.	In all stands that are underplanted with white pine and in most stands with natural white pine regeneration, treat seedlings annually with deer repellent until they are approximately 6 feet tall.	Immediate, short-term (10 years or less)	• This is already done in most stands. • Higher likelihood of success in regenerating white pine and other desirable forest species. • Overall higher-quality ecosystem expected to result.	• Labor and other costs	High	Yes

106

Worksheet #4/Step #4 (continued): IDENTIFY adaptation approaches and tactics for implementation.

Adaptation Approach	Tactic	Time Frames	Benefits	Drawbacks and Barriers	Practicability of Tactic	Recommend Tactic?
Manage herbivory to protect or promote regeneration.	In selected or high-priority stands with paper birch regeneration (natural or planted), treat seedlings annually with deer repellent until they are approximately 6 feet tall. Select stands based on access, site conditions, aesthetics, and/or ease for monitoring.	Immediate, short-term, medium-term (30 years or less). Experimental.	• This is frequently done for white pine and other species, but not for paper birch. • Higher likelihood of success in regenerating paper birch and other desirable forest species. • Overall higher-quality ecosystem expected to result.	• Use of deer repellent spray is experimental for paper birch, and it is uncertain whether regeneration will be enhanced. • Labor and other costs may be high. • Annual application of repellent will need to be tracked, and success or failure will need to be evaluated.	Low	Yes: Experimental
	Use fencing to exclude deer from selected or high-priority paper birch stands. Select stands based on access, site conditions, aesthetics, and/or ease for monitoring.	Immediate, short-term, medium-term (30 years or less). Experimental.	• Fencing is often very effective at reducing deer browse. • Successful regeneration of paper birch and other desirable forest species is expected to be more likely. • Overall higher-quality ecosystem is expected to result.	• Labor and other costs may be high. • Success or failure will need to be evaluated.	Moderate	Yes: Experimental
Alter forest structure to reduce severity or extent of wind and ice damage.	Refer to above tactics. (Regeneration of the currently over-mature stands will result in younger stands that will be less vulnerable to wind and ice damage in the short term and medium term.)					
Promote diverse age classes.	Refer to above tactics. (Regeneration of the stands will diversify forest structure over the long term and help achieve age class distribution goals. Across the broader project area, a patchwork of newly harvested and older unharvested stands will increase landscape heterogeneity. Within these stands that are being regenerated in the near future, retaining biological legacies and favoring white pine on many sites will increase the structural diversity within stands.)					
Maintain and restore diversity of native tree species.	Refer to above tactics. (Regeneration of the over-mature paper birch stands will help to retain native tree species and early-successional habitat that are important to this area.)					

(Worksheet #4 continued on next page.)

Worksheet #4/Step #4 (continued): IDENTIFY adaptation approaches and tactics for implementation.

Adaptation Approach	Tactic	Time Frames	Benefits	Drawbacks and Barriers	Practicability of Tactic	Recommend Tactic?
Maintain and restore diversity of native tree species.	Retain native trees that are: (1) currently under-represented in stands or (2) expected to be better adapted to future conditions. Favor species for retention that do not compete with paper birch regeneration. Desirable species include red, white, and bur oak; basswood; white ash; and red, white, and jack pine. Retained trees will diversify stands to reduce overall risk of decline and provide a future and continued seed source. These individuals will also serve as reserve trees in the short term and woody debris in the long term.	Immediate, short-term, medium-term, long-term	• Forest Plan goals and objectives for reserve trees, species diversity, coarse woody debris, under-represented species, and more are addressed. • Retention of under-represented tree species is commonly done throughout the forest. This tactic would build on those efforts. • Retention of a diversity of tree species may offset expected declines in some conifer species (e.g., balsam fir) by encouraging or promoting less vulnerable conifer species (e.g., red pine, white pine).	• Some stands do not have many tree species or other biological legacies to begin with, so there may not be much available to work with in these areas.	Extremely high (especially in stands that have biological legacies).	Yes
	After treating selected stands, underplant low densities of northern red, white, and bur oak on a small number of sites to increase diversity and encourage species that are likely better adapted to future conditions. Select stands based on such criteria as Ecological Land Type Phase and site conditions. The underplanting of white and bur oak would be experimental.	Immediate, short-term (10 years or less)	• Actively diversifies stands with native oak species that are expected to have increased habitat suitability in the future.	• Is labor-intensive to plant and protect young seedlings.	Moderate	Yes
Retain biological legacies.	Refer to above tactics. (The retention of trees that are currently under-represented or that are likely to be adapted to future conditions will help provide biological legacies.)					

Worksheet #4/Step #4 (continued): IDENTIFY adaptation approaches and tactics for implementation.

Adaptation Approach	Tactic	Time Frames	Benefits	Drawbacks and Barriers	Practicability of Tactic	Recommend Tactic?
Retain biological legacies.	Implement Forest Plan retention guidelines.	Immediate (2 years or less)	• Addresses Forest Plan goals and objectives for reserve trees, species diversity, coarse woody debris, under-represented species, and more. • Is already occurring throughout the forest.		Extremely high	Yes
Restore fire to fire-adapted ecosystems.	Select sites to shelterwood followed by prescribed burning for paper birch regeneration based on stand shape, accessibility, topography, and adjacent landownership. Look for opportunities to use prescribed burning over mechanical scarification. For example: (1) include adjacent stands that are not paper birch in prescribed burns; (2) locate groups of stands that would benefit from prescribed burning and where the burning is logistically feasible; or (3) collaborate with fuel reduction planning efforts.	Immediate (2 years or less)	• Enhanced regeneration is expected in these areas because paper birch responds well to fire. • Less soil disturbance is expected because little or no mechanical scarification is likely to be needed where burns occur.	• Social acceptance to burning can be low. • Prescribed burn windows are often not available for the very hot type of burn that is needed to prepare the seedbed. • There are many logistical challenges to conducting burns where the paper birch stands are located (e.g., near private lands, low accessibility, challenging topography, high amounts of fireline construction needed).	Low	Yes
Manage habitats over a range of sites and conditions.	Refer to above tactics. (Regeneration of stands over a range of sites is planned to occur where possible. Additional efforts to identify and maintain refugia in the long term will also help to maintain habitats across a variety of sites and conditions.)					

(Worksheet #4 continued on next page.)

APPENDIX 4. PAPER BIRCH FOREST ILLUSTRATION; STEP #4

Worksheet #4/Step #4 (continued): IDENTIFY adaptation approaches and tactics for implementation.

Adaptation Approach	Tactic	Time Frames	Benefits	Drawbacks and Barriers	Practicability of Tactic	Recommend Tactic?
Use seeds, germplasm, and other genetic material from across a greater geographic range.	For stands where white pine is underplanted, purchase stock from inside and outside of the immediate area (e.g., from farther south, east, or west). Keep records of what was used at different locations and track results of different stock over time.	Immediate, short-term (10 years or less)	• Using a wider variety of sources for planting stock would increase the likelihood of introducing genotypes that are better adapted to future conditions. • It may be easier to maintain white pine on the landscape in the future.	• The type and amount of growing stock suited to warmer and drier conditions may be limited. • Some genotypes may be less well-adapted to current or future conditions. • Current guidelines in the Forest Plan recommend using stock from within the same climatic zone.	Moderate	Yes
Favor existing genotypes that are better adapted to future conditions.	For areas where paper birch is artificially regenerated, purchase seeds and/or planting stock from sources in warmer climatic zones.	Immediate, short-term, medium-term (30 years or less)	• Using seed or planting stock that is expected to be better adapted to future conditions would increase the likelihood of maintaining paper birch on the landscape in the long term.	• Since paper birch is not normally planted or seeded, it may be difficult to find sources of genetic material. • Paper birch is near the southern extent of its range in northern Wisconsin, so the number of sources adapted to warmer and drier conditions may be limited. • Additional effort will be needed to seed or plant the genetic material successfully.	Short-term: Low Long-term: High	Yes
Increase diversity of nursery stock to provide those species or genotypes likely to succeed.	Develop hybrids and/or improved paper birch genotypes better adapted to warmer and drier climates.	Long-term (30 years or more)	• Using hybrids or improved genotypes that are expected to be better adapted to future conditions would increase the likelihood of maintaining desired species on the landscape in the long term.	• The development of these hybrids and improved genotypes is beyond the Forest's capability, and research partners would be needed. • It would be important to think ahead to obtain stock that is resistant to diseases and better adapted to future conditions.	Short-term: Low Long-term: High	Yes

APPENDIX 4. PAPER BIRCH FOREST ILLUSTRATION; STEP #4

Worksheet #4/Step #4 (continued): IDENTIFY adaptation approaches and tactics for implementation.

Adaptation Approach	Tactic	Time Frames	Benefits	Drawbacks and Barriers	Practicability of Tactic	Recommend Tactic?
Anticipate and respond to species decline.	Refer to above tactics. (The combination of many tactics helps to respond to the predicted declines of many important species.)					
Favor or restore native species that are expected to be better adapted to future conditions.	Refer to above tactics. (Many efforts in the short term will focus on maintaining a diversity of native tree species. Having diverse stands will provide a range of options and allow managers to determine which species are better adapted to conditions at individual locations.)					
	Encourage conversion of paper birch to northern hardwood, oak, white pine, or other forest types on sites where repeated regeneration efforts are unsuccessful.	Medium-term, long-term (in ~10 or more years)	• Given the expected decline in paper birch habitat, it may not be feasible to keep paper birch on all sites. Allowing stands that cannot be managed as paper birch to convert to other forest types will retain forest cover and forest habitats into the future.	• In stands that are converted to other types, the paper birch component will be significantly reduced or even lost.	Moderate	Yes
Manage species and genotypes with wide ranges of moisture and temperature tolerances.	Refer to above tactics. (White pine has a wide ecological amplitude, which makes it more likely than many native species to persist under a range of future conditions.)					
	Encourage conversion of paper birch to mixed hardwoods (which includes maple species) where repeated paper birch regeneration efforts are unsuccessful.	Medium-term, Long-term (in ~10 or more years)	• Given the expected decline in paper birch habitat, it may not be feasible to keep paper birch on all sites. Allowing stands that cannot be managed as paper birch to convert to other forest types will retain forest cover and forest habitats into the future. • Mixed hardwood stands may be able to persist under a range of conditions because of the diversity of tree species present in these stands as well as red maple's ability to thrive across a wide variety of sites.	• In stands that are converted to other types, the paper birch component will be significantly reduced or even lost.	Moderate	Yes
Emphasize drought- and heat-tolerant species and populations.	Refer to above tactics. (Many efforts in the short term will focus on maintaining a diversity of native tree species. Having diverse stands will provide a range of options and allow managers to determine which species are better adapted to conditions at individual locations.)					

(Worksheet #4 continued on next page.)

Worksheet #4/Step #4 (continued): IDENTIFY adaptation approaches and tactics for implementation.

Adaptation Approach	Tactic	Time Frames	Benefits	Drawbacks and Barriers	Practicability of Tactic	Recommend Tactic?
Guide species composition at early stages of stand development.	Refer to above tactics. (Many of the tactics focus on regeneration of paper birch stands that are currently over-mature to help guide species composition for the next rotation and beyond.)					
Protect future-adapted regeneration from herbivory.	Refer to above tactics. (A number of tactics exist to counter deer herbivory, and these efforts can be expanded to protect other species as needed.)					
Establish or encourage new mixes of native species.	In paper birch stands where repeated paper birch regeneration efforts are unsuccessful, consider planting different species mixes, such as red, white, and/or bur oak.	Immediate, short-term, medium-term (30 years or less)	• New species mixes would increase stand diversity and provide a range of options for the future. • There is a greater chance that the future stand would be resilient to pest and disease outbreaks because of the diversity of species. • The failure of paper birch regeneration in some stands would provide an opportunity to try new approaches.	• This is a new practice, so there is a steep learning curve. Managers are uncertain about which species would be best to use because the future site conditions and suitable species are not known. • It may be hard to defend this action depending on the species that were used.	Moderate	Yes
Prepare for more frequent and more severe disturbances.	Use Forest Plan procedures for responding to large areas of natural disturbance.	Immediate (2 years or less) following large disturbance	• Guidance already exists for how to respond to disturbances based on management area objectives. • Response actions can be expedited for disturbances under 250 acres.	• It is not possible to plan in detail prior to disturbance because the response is highly dependent upon the type and severity of the disturbance, the condition of the site following disturbance, and the management goals and opportunities for the site.	Extremely high	Yes

112

Worksheet #4/Step #4 (continued): IDENTIFY adaptation approaches and tactics for implementation.

Adaptation Approach	Tactic	Time Frames	Benefits	Drawbacks and Barriers	Practicability of Tactic	Recommend Tactic?
Allow for areas of natural regeneration after disturbance.	Use Forest Plan for guidance for responding to large areas of natural disturbance. At least 5-15 percent of the site should be left untreated following disturbance based on the management area goals.	Immediate (2 years or less) following large disturbance	• Guidance already exists for how to respond to disturbances based on management area objectives.		Extremely high	Yes
Maintain seed or nursery stock of desired species for use following severe disturbance.	Collect seeds of a variety of tree species, including paper birch, white pine, and oaks. Coordinate with a nursery regarding storage.	Medium-term (10-30 years)	• Obtaining seeds now may provide opportunities to plant native species in the future.	• Resources will be needed to collect and store seeds, and a nursery partner will be needed.	Moderate	Yes
Remove or prevent establishment of invasives and other competitors following disturbance.	Refer to above tactics. (A number of tactics exist to counter invasive species, and these efforts can be expanded to newly disturbed areas as needed.)					
	Introduce fire into newly disturbed areas when desirable and feasible, to prevent raspberry and other invasive species from becoming established.	Immediate (2 years or less) following large disturbance	• The use of prescribed burning may enhance the regeneration of paper birch and other desired species. • Less soil disturbance is expected because little or no mechanical scarification is likely to be needed where burns occur.	• Social acceptance to burning can be low. • There are many logistical challenges to conducting burns, and these challenges may be greater in areas where there is blowdown or other disturbance.	Low	Yes
Promptly revegetate sites after disturbance.	Use Forest Plan for guidance for responding to large areas of natural disturbance.	Immediate (2 years or less) following large disturbance	• Guidance already exists for how to respond to disturbances based on management area objectives.		Extremely high	Yes

(Worksheet #4 continued on next page.)

113

Worksheet #4/Step #4 (continued): IDENTIFY adaptation approaches and tactics for implementation.

Adaptation Approach	Tactic	Time Frames	Benefits	Drawbacks and Barriers	Practicability of Tactic	Recommend Tactic?
Promptly revegetate sites after disturbance.	Where there is an adequate overstory remaining after blowdown event, conduct prescribed burning or mechanical scarification for natural paper birch regeneration.	Immediate (2 years or less) following large disturbance	• A disturbance may provide an opportunity to establish and manage paper birch. • Younger birches tend to be less vulnerable and more resilient to many stressors. • Regenerating paper birch helps meet goals and objectives set out in the Forest Plan. • The use of prescribed burning may enhance the regeneration of paper birch and other desired species.	• Regeneration is not guaranteed after treatment. • It is uncertain whether this approach reduces the long-term vulnerability of paper birch to climate change. • Success is often dependent upon site conditions. • Paper birch regeneration is difficult because a large amount of site preparation is needed. • Sites suitable for paper birch regeneration treatments are often limited because of factors such as accessibility, aesthetic concerns, and topographic limitations.	Short-term: Moderate Long-term: Low	Yes
	Take advantage of existing advanced regeneration where present (e.g., hardwoods, white pine, aspen).	Immediate (2 years or less) following large disturbance	• A disturbance may provide an opportunity to establish and manage a variety of tree species. • Retention of under-represented tree species is commonly done throughout the forest. This tactic would build on those efforts. • Retention of a diversity of tree species would help to offset the loss of trees due to the disturbance.		Extremely high	Yes

114

Worksheet #5/Step #5: MONITOR and evaluate effectiveness of implemented actions.

Monitoring Items	Monitoring Metric(s)	Criteria for Evaluation	Monitoring Implementation
Objective 1: Regenerate the existing mature paper birch to retain it on the landscape when desirable.	• Number of acres regenerated	• Passes stocking survey.	• Monitor seedling success during 3rd- and 5th-year stocking survey. If a stand fails, update monitoring for follow-up activity.
Objective 2: Regenerate or underplant white pine among the natural paper birch regeneration when opportunities exist to improve stand diversity, when paper birch regeneration is not possible, or when site scarification is not possible or desired.	• Number of acres underplanted or regenerated	• Passes stocking survey.	• Monitor seedling success during 1st- and 3rd-year stocking survey for planted species. If a stand fails, update monitoring for follow-up activity.
Multiple tactics related to regeneration of paper birch, white pine, and other species: 1) Treat selected over-mature paper birch stands with a shelterwood harvest followed by prescribed burning or mechanical site preparation. Prioritize the stands to be treated using a field check of site conditions. 2) Treat selected over-mature paper birch stands with a shelterwood harvest followed by Bracke scarification and underplanting of white pine. Leave the overstory. 3) Select sites to shelterwood followed by prescribed burning for paper birch regeneration based on stand shape, accessibility, topography, and adjacent landownership. Look for opportunities to use prescribed burning over mechanical scarification. For example: (1) include adjacent stands that are not paper birch in prescribed burns; (2) locate groups of stands that would benefit from prescribed burning and where the burning is logistically feasible; or (3) collaborate with fuel reduction planning efforts. 4) Treat selected over-mature paper birch stands that have an existing white pine seed source or advance regeneration with a shelterwood harvest and scarify for white pine. Underplant white pine to augment advance regeneration if needed. Retain the overstory. 5) After treating selected stands, underplant low densities of northern red, white, and bur oak on a small number of sites to increase diversity and encourage species that are likely better adapted to future conditions. Select stands using criteria such as Ecological Land Type Phase and site conditions. The underplanting of white and bur oak would be experimental.	• Treatment acres ÷ over-mature acres • Number of acres prescribed burned • Number of acres mechanically scarified • Number of acres regenerated • Number of acres underplanted	• Passes stocking survey. • Compare prescribed burn results to mechanical scarification to evaluate paper birch regeneration success.	• Review National Environmental Policy Act decision to determine number of over-mature acres that were treated. • Monitor seedling success during 1st-, 3rd-, and 5th-year stocking surveys. Identify needed monitoring for any follow-up activities.

(Worksheet #5 continued on next page.)

Worksheet #5/Step #5 (continued): MONITOR and evaluate effectiveness of implemented actions.

Monitoring Items	Monitoring Metric(s)	Criteria for Evaluation	Monitoring Implementation
Regenerate paper birch and establish white pine on a range of site conditions.	• Number of stands or acres of each type regenerated and/or maintained on different Ecological Land Type Phases	• To be determined	• Query FSVeg database (or other database) at 5-year intervals.
Adjust rotation age lengths to achieve age class distribution goals described in the Land and Resource Management Plan.	• Age class distribution before and after treatment	• Forest Plan goals for age class distribution	• Use FSVeg database.
Use the recognized NNIS introduction and spread tactics in the NNIS guide, contract provisions, and other applicable sources (e.g., equipment cleaning requirements, weed-free seed).	• Implementation of NNIS tactics and guidelines (Yes/No)	• Refer to existing Forest-wide NNIS monitoring criteria.	• Refer to existing Forest-wide NNIS monitoring protocol.
Use stocking surveys and NNIS surveys to determine whether regeneration is being adversely affected by native and nonnative invasive plant species. If it is, consider herbicide treatments of the invasives, additional scarification, and/or individual tree releases.	• Number of acres requiring follow-up treatment	• Passes stocking survey • Certification of release	• Monitor seedling success using stocking surveys. • Identify reasons for regeneration failures (e.g., competing vegetation). • Identify needed monitoring for any follow-up activities.
Ensure adequate overstory is retained so as not to encourage establishment of sun-loving invasives.	• Basal area or percent canopy closure	• To be determined. Approximate values: 40-60 ft^2 basal area or >50 percent canopy closure	• Post-marking and/or post-harvest evaluation
Multiple tactics related to reducing deer herbivory: 1) In all stands that are underplanted with white pine and in most stands with natural white pine regeneration, treat seedlings annually with deer repellent until they are approximately 6 feet tall. 2) In selected or high-priority stands with paper birch regeneration (natural or planted), treat seedlings annually with deer repellent until they are approximately 6 feet tall. Select stands based on access, site-conditions, aesthetics, and/or ease for monitoring.	• Implementation of treatment (Yes/No) • Number of acres treated • Percent of trees treated (if in contract specifications)	• 100 percent of planted white pine stands are treated. • A given percentage of individuals in unplanted white pine stands are treated—value to be determined. • 100 percent of selected paper birch stands are treated. • Regeneration is more effective in these stands than in similar paper birch stands not treated with repellent or fenced.	• Silviculturist makes an annual assessment (similar to stocking survey). • Use FACTS database. • Persons responsible for application of deer repellent also perform monitoring to determine whether an adequate percentage of trees were treated.

Worksheet #5/Step #5 (continued): MONITOR and evaluate effectiveness of implemented actions.

Monitoring Items	Monitoring Metric(s)	Criteria for Evaluation	Monitoring Implementation
Use fencing to exclude deer from selected or high-priority paper birch stands. Select stands based on access, site-conditions, aesthetics, and/or ease for monitoring.	• Installation of fencing (Yes/No). • Number of acres and/or sites fenced	• 100 percent of selected stands are fenced. • Regeneration is more effective in these stands than in similar paper birch stands not treated with repellent or fenced.	• Monitor seedling success using stocking surveys. • Identify needed monitoring for any follow-up activities.
Evaluate the stands that successfully regenerate, and identify some stands to serve as future refugia for paper birch.	• Number of acres identified	• To be determined	• To be determined
When selecting stands for paper birch regeneration, look for sites that may be suitable in the future. For example, emphasize stands on cooler and moister sites, such as north aspects, bog islands, and eskers within lower-elevation areas.	• Number of stands identified	• To be determined	• To be determined
Retain native trees that are: (1) currently under-represented in stands or (2) expected to be better adapted to future conditions. Favor for retention species that do not compete with paper birch regeneration, such as red, white, and bur oak; basswood; white ash; and red, white, and jack pine. Retained trees will diversify stands to reduce overall risk of decline and provide a future and continued seed source. They will also serve as reserve trees in the short term and woody debris in the long term.	• Trees were retained where possible (Yes/No). • Additional metrics will to be determined (such as number of trees and areas reserved, number of different species reserved, or stand diversity index).	• Met Forest Plan goals for reserve tree guidelines by management area and forest type. • Trees retained on-site are consistent with what was prescribed.	• Use timber sale reviews and stand exams to compare silvicultural prescription with implementation. • Managers also may need to develop a way to monitor minor stand component changes over time.
Multiple tactics related to seedlings and seed sources: 1) For stands where white pine is underplanted, purchase stock from inside and outside of the immediate area (e.g., from farther south, east, or west). Keep records of what was used at different locations and track results of different stock over time. 2) For areas where paper birch is artificially regenerated, purchase seeds and/or planting stock from sources in warmer climatic zones.	• Number of trees planted using different sources and/or locations • Number of sources used • Survival rate by source	• Passes stocking survey (short-term) • Passes a survey of long-term survival and condition	• Monitor seedling success using stocking surveys. Identify needed monitoring for any follow-up activities. • Coordinate with research partners for long-term evaluation of stock from alternative sources/locations
Develop hybrids and/or improved paper birch genotypes better adapted to warmer and drier climates.	• n/a	• n/a	• Coordinate with research partners.

(Worksheet #5 continued on next page.)

Worksheet #5/Step #5 (continued): MONITOR and evaluate effectiveness of implemented actions.

Monitoring Items	Monitoring Metric(s)	Criteria for Evaluation	Monitoring Implementation
Multiple tactics related to conversion of paper birch stands that fail to regenerate: 1) Encourage conversion of paper birch to northern hardwood, oak, white pine, or other forest types on sites where repeated regeneration efforts are unsuccessful. 2) Encourage conversion of paper birch to mixed hardwoods (which includes maple species) where repeated paper birch regeneration efforts are unsuccessful. 3) In paper birch stands where repeated paper birch regeneration efforts are unsuccessful, consider planting different species mixes, which could include red, white, and or bur oak.	• Number of sites or acres where initial regeneration efforts were unsuccessful • Percentage of unsuccessful sites or acres converted to other species or forest types	• Passes stocking surveys	• Monitor seedling success using stocking surveys. • Identify needed monitoring for any follow-up activities.
Multiple tactics related to response to disturbance: 1) Use Forest Plan for guidance for responding to large areas of natural disturbance 2) Use CNNF procedures that outline actions for responding to large areas of natural disturbance.	• Use of guidance (Yes/No) • Number of acres disturbed • Number of acres treated • Number of acres left untreated	• All disturbed acres were included in a response to the disturbance	• To be determined.
Use Forest Plan for guidance for responding to large areas of natural disturbance. At least 5-15 percent of the site should be left untreated following disturbance based on the management area goals.	• Number of acres left untreated	• At least 5-15 percent of acres are left untreated per event unless more restrictive criteria apply (Forest Plan guideline).	• Review NEPA decision.
Collect seeds of a variety of tree species, including paper birch, white pine, and oaks. Coordinate with a nursery regarding storage.	• Collection of seeds (Yes/No) • Amount of seeds collected • Number of species collected	• To be determined	• To be determined. Coordinate with nurseries.
Introduce fire into newly disturbed areas when desirable and feasible, to prevent raspberry species and other invasives from becoming established.	• Number of sites or acres treated	• Compare prescribed burn results to mechanical scarification to evaluate paper birch regeneration success. • Refer to existing Forest-wide NNIS monitoring criteria	• Monitor seedling success using stocking surveys. • Identify reasons for any regeneration failures (e.g., competing vegetation).

Worksheet #5/Step #5 (continued): MONITOR and evaluate effectiveness of implemented actions.

Monitoring Items	Monitoring Metric(s)	Criteria for Evaluation	Monitoring Implementation
Where there is an adequate overstory remaining after blowdown event, conduct prescribed burning or mechanical scarification for natural paper birch regeneration.	• Number of acres of blowdown area that are prescribed burned	• Compare prescribed burn results to mechanical scarification to evaluate paper birch regeneration success.	• Monitor seedling success using stocking surveys. • Identify needed monitoring for any follow-up activities.
Take advantage of existing advanced regeneration where present (e.g., hardwoods, white pine, aspen).	• Retention of trees where possible and where paper birch regeneration would not be negatively impacted (Yes/No) • Additional metrics to be determined (such as number of trees and areas reserved, number of different species reserved, or stand diversity index).	• Forest Plan goals for reserve tree guidelines were met by management area and forest type. • Trees retained on-site are consistent with what was prescribed.	• Use timber sale reviews and stand exams to compare silvicultural prescription with implementation. • Managers also may need to develop a way to monitor minor stand component changes over time.

ADAPTATION WORKSHEETS

Worksheet #1

Area of Interest	Location	Forest Type(s)	Management Goals	Management Objectives	Time Frames

Worksheet #2

Broad-scale Impacts and Vulnerabilities	Climate Change Impacts and Vulnerabilities for the Area of Interest	Vulnerability Determination
	How might broad-scale impacts and vulnerabilities be affected by conditions in your area of interest? • Landscape pattern • Site location, such as topographic position or proximity to water features • Soil characteristics • Management history or current management plans • Species or structural composition • Presence of or susceptibility to pests, disease, or nonnative species that may become more problematic under future climate conditions • Other….	

Worksheet #3

Management Objective (from Worksheet #1, column 5)	Challenges to Meeting Management Objective with Climate Change	Opportunities for Meeting Management Objective with Climate Change	Feasibility of Meeting Objective under Current Management	Other Considerations

Worksheet #4

Adaptation Approach	Tactic	Time Frames	Benefits	Drawbacks and Barriers	Practicability of Tactic	Recommend Tactic?

Worksheet #5

Monitoring Items	Monitoring Metric(s)	Criteria for Evaluation	Monitoring Implementation

Swanston, Chris; Janowiak, Maria, eds. 2012. **Forest adaptation resources: Climate change tools and approaches for land managers.** Gen. Tech. Rep. NRS-87. Newtown Square, PA: U.S. Department of Agriculture, Forest Service, Northern Research Station.121 p.

The forests of northern Wisconsin, a defining feature of the region's landscape, are expected to undergo numerous changes in response to the changing climate. This document provides a collection of resources designed to help forest managers incorporate climate change considerations into management and devise adaptation tactics. It was developed in northern Wisconsin as part of the Northwoods Climate Change Response Framework project and contains information from assessments, partnership efforts, workshops, and collaborative work between scientists and managers. The four interrelated chapters include: (1) a description of the overarching Climate Change Response Framework, a landscape-scale conservation approach also being expanded to other landscapes; (2) a "menu" of adaptation strategies and approaches that are directly relevant to forests in northern Wisconsin; (3) a workbook process to help incorporate climate change considerations into forest management planning and to assist land managers in developing ground-level climate adaptation tactics for forest ecosystems; and (4) two illustrations that provide examples of how these resources can be used in real-world situations. The ideas, tools, and resources presented in the different chapters are intended to inform and support the existing decisionmaking processes of multiple organizations with diverse management goals.

KEY WORDS: Adaptation, climate change, Wisconsin, forest, adaptive management, Midwest, landscape, case study